DON'T KNOW

MUCH ABOUT

GEOGRAPHY

DON'T KNOW

MUCH ABOUT

GEOGRAPHY

Everything You Need

to Know About the

World but Never Learned

KENNETH C. DAVIS

HARPER

NEW YORK ● LONDON ● TORONTO ● SYDNEY

HARPER

Don't Know Much About® is a registered trademark of Kenneth C. Davis.

A hardcover edition of this book was published in 1992 by William Morrow, an imprint of HarperCollins Publishers.

HarperCollins books may be purchased for educational, business, or sales promotional use. For information please e-mail the Special Markets Department at SPsales@harpercollins.com.

First Avon Books edition published 1993.
Reissued in Harper Perennial 2001, 2004, 2013.

The Library of Congress has catalogued the hardcover edition of this book as follows:

Davis, Kenneth C.
   Don't know much about® geography: everything you need to know about the world but never learned / Kenneth C. Davis
         p. cm.
   Includes bibliographical references and index.
            1. Geography—Miscellanea. 2. Questions and answers. I. Title.
G131.D38 1992   92-19142
910'.76—dc20

ISBN 978-0-06-204356-6 (pbk.)

17   OV/RRD   10 9 8 7 6 5 4 3 2

To my parents,
Evelyn and Richard Davis,
who gave me good directions.

Geography is a representation of the whole known world together with the phenomena which are contained therein.

In Geography one must contemplate the extent of the entire Earth, as well as its shape, and its position under the heavens . . . the length of its days and nights, the stars which are fixed overhead, the stars which move above the horizon, and the stars which never rise above the horizon at all. . . .

It is the great and exquisite accomplishment of mathematics to show all these things to human intelligence.

—PTOLEMY
GEOGRAPHIA

No two countries that both had McDonald's had ever fought a war against each other since each got its McDonald's. (Border skirmishes and civil wars don't count since McDonald's usually served both sides.) . . . The Golden Arches Theory of Conflict Prevention . . . When a country reached the level of economic development where it had a middle class big enough to support a network of McDonald's it became a McDonald's country. And people in McDonald's countries didn't like to fight wars anymore.

—THOMAS L. FRIEDMAN
THE WORLD IS FLAT (2007)

What kind of world is likely if we take no deliberate action?
What kind of world do we want?
What kind of world is possible if we act effectively?

—WILLIAM HOOKE
WWW.LIVINGONTHEREALWORLD.ORG

Contents

Preface to the Revised Edition of
Don't Know Much About® Geography

Okay. They always tell you to lead with the headline. So here goes. BREAKING NEWS: They've added an ocean.

Say what?

That's right. There are five oceans now where there used to be only four. Or, at least, so says the IHO (no, not IHOP). The International Hydrographic Organization is a multinational, intergovernmental group that oversees issues of navigation and charting and other oceanic matters. Based in Monaco—I know; tough job, but somebody's gotta do it— the IHO decided a few years ago that there is a Southern Ocean, or as some call it, an Antarctic Ocean. As readers of the original book would know, this is not a brand new idea. I discussed the subject of the Southern Ocean in the first edition of the book. But now it's more "official."

This neo-ocean extends up from Antarctica and would count as the world's fourth largest ocean—if it counted. This is in addition to the four we were supposed to learn back in the fifth grade. (The Pacific, Atlantic, Indian, and Arctic, in size order, just in case you were absent or just bored that day.)

Now you may also have heard that Pluto was downgraded from planetary to "dwarf planet" status. There's still some dispute about that one, too, but it did get plenty of media attention.

But coming up with a new ocean! You would think that would have made the front pages.

But this Earth- (or ocean-) shaking news hasn't exactly taken the world by storm. Many people and reference books, including the *Time Almanac 2012*, do not seem to know about this extra ocean. Or not everyone recognizes its existence. The Rodney Dangerfield of oceans—it gets no respect.

It must also be said that there is in fact only one ocean—the great single body of interconnected water that covers much of the earth's surface and is only occasionally broken up by the little plots of land called continents and islands.

And that is one reason why geography can be so confusing, as I wrote in this book when it was originally published. Geography, from the Greek meaning "to describe the earth," is sometimes as much art as science and some terms are less easily defined. Seas can be lakes. Jungles can be rain forests. And continents can be islands.

In the two decades since this book was first written, the earth hasn't changed much. Oh sure, along with that new ocean, we've added millions of people to the head count—with billions more expected to arrive on board Spaceship Earth over the next few decades. Some of the borders have changed; countries have broken apart—like Czechoslovakia, which peacefully split into the Czech Republic and Slovakia in January 1993. Earthquakes, volcanoes, floods, and erosion continue to reshape the earth's features. And the global temperature continues to climb.

When this book first appeared, America had just ended a war against Iraq led by President Bush. Now in 2012, America has just ended a war against Iraq begun by President Bush. Of course, different wars, different presidents Bush, and very different outcomes.

So while some things stay the same, there have been enormous changes in the world in twenty years. And this book has been updated and revised with new questions to reflect these essential changes:

- The role of the Internet and other technology in transforming global life (*What was the Arab Spring?*)

- The rise of China, India and other former "developing nations" as world economic powers amid the globalization of commerce (*What can you build with BRICS? Does the World Bank have ATMs?*)

- The question of sustainability in a world that is growing, as author Thomas Friedman succinctly put it, "hot, flat, and crowded"

- The debate over climate change and evolutionary science and the related question of how science has become a partisan political issue—particularly in the United States (*Is all the talk of global warming just a lot of hot air?*)

As I wrote in the original Introduction to this book, geography is not just about memorizing place names and state capitals or knowing how to read maps. It is about understanding our place in the world and who our neighbors are. It is about understanding the links between places and events. My goal was to get people to "think geographically"—to look at the world with the great sense of curiosity that some of the ancient thinkers possessed—and attempt to figure out the world. That was the beginning of science. Which raises the most serious point of this book.

During the past twenty years, America has witnessed a concerted assault on science. While some of those attacks come from the fact that we learn new things about medicine, space, and biology all the time—and yes, we know that science and scientists can be wrong—much of the assault has come from people with very specific agendas. Those agendas can be motivated by profit, political ideology, religious belief, or faith in what has been called "junk science."

But the serious threat to good science and, more importantly, science education is a dangerous thing, especially in a world that will increasingly demand complex technological and scientific answers to its pantheon of problems. I have tried to address some of these hot-button issues, especially climate change and evolution, very directly.

As I write this, a 2012 Gallup survey showed that 46 percent of Americans believe in the creationist view that God created humans in their present form at some time within the last ten thousand years.* That belief, largely a matter of faith among some Christians, has been aggressively introduced into American public education in recent years. Once known as "creationism," this idea was repackaged as "intelligent design," or ID—an attempt to put very old wine into a new bottle. In a closely watched court case in Pennsylvania, the intelligent design movement made a thinly veiled attempt to question all of evolution-

* www.gallup.com/poll/155003/Hold-Creationist-View-Human-Origins.aspx

ary biology by introducing doubt over relatively small and unresolved issues. This was an end-run approach to introduce the biblical view of creation, previously ruled unconstitutional by the Supreme Court, into science classes. The strategy was completely rejected in December 2005 by U.S. District Judge John E. Jones III, who wrote in his ruling, "[We] find that ID [Intelligent Design] is not science and cannot be adjudged a valid, accepted scientific theory as it has failed to publish in peer-reviewed journals, engage in research and testing, and gain acceptance in the scientific community. ID, as noted, is grounded in theology, not science. Accepting for the sake of argument its proponents', as well as Defendants' argument that to introduce ID to students will encourage critical thinking, it still has utterly no place in a science curriculum."*

That ruling by a Republican administration-appointed judge dismantled the ID argument. Yet the issue—and the fundamental belief attacking evolutionary science behind it—has not gone away.

It is my hope that this book will shed more light than heat about geography and its wonders.

Looking back at the changes over the past two decades, both historically and technologically, it is difficult to imagine writing about the world twenty years from now. But geography helps by showing us where we have been and, maybe, where we are going.

Now, about those oceans: When the question "How many oceans are there?" comes up and the multiple-choice answers are:

a. One
b. Four
c. Five
d. All of the above

Now you know. Go with D.

*www.nytimes.com/2005/12/21/education/21evolution.html?pagewanted=all

INTRODUCTION
How Come the Nile River Flows Up?

Way back in elementary school, I had a social studies teacher I'll call Mrs. McNally. One day, in the middle of a geography lesson, Mrs. McNally lost it. Things started to fall apart for her when she pulled down one of those wonderful window-shade maps we had in grade school.

Remember them? Three or four maps mounted over the blackboard? You pulled one down and it usually snapped right back up again. Geography class sometimes looked like a Three Stooges routine. (Yes, children, once upon a time, we had maps on paper, not on laptops, PCs, smartphones, and tablets.)

On this particular morning, it was a map of Africa because the class was studying Egypt and the Nile River. As the teacher spoke, a small hand shot up and a tiny voice asked, "How come the Nile River flows up?"

Today, it seems like a silly, yet innocent, child's question. (If you are asking yourself the same question, then you really need this book!) But back then, it was a puzzle that immediately caught the attention of the whole class. With the proverbial light bulb clicking on over our heads, we all wondered, "Yeah, how could a river flow up?"

To our fifth-grade minds, it simply made no sense for a river to flow *up the map*. Everybody knew that water had to flow down. We had caught the teacher in a very obvious mistake.

Of course, on a map oriented along the lines of this jingle:

North to the ceiling,
South to the floor,
West to the window,
East to the door

it did appear that the Nile River flowed up.

I can't tell you much else about what happened in that classroom that year. But I can report that this question prompted a small classroom revolt. Try as she might, Mrs. McNally could not get across to this group of ten-year-olds that there was a difference between *up* and the compass direction *north* depicted on the map, and that consequently the Nile River actually flowed down from the mountains in East Africa to the Mediterranean Sea.

I don't remember exactly how she tried to make this point, but I do recall that she failed—miserably. It was hopeless. Mrs. McNally grew so frustrated with our inability to grasp this elementary geographic concept that she blew her top. We ended up receiving some ghastly—and unjust—punishment, like no recess or a day's detention for the entire class.

Through the years, that nightmarish geography lesson has stayed with me. I suppose it's a grim example of how grossly inadequate teachers can sometimes be. (Before the teachers' union comes after me with a noose, I'll add that I have had a great many wonderful teachers who gave me a love of learning and who made the classroom a pleasure. They deserve much more credit than they get from educational critics.)

But that lesson also leaves me wondering how many kids actually got things straight. I'm sure that Mrs. McNally wasn't the only one out there struggling to get a point across. Her failure as a geography teacher, it now seems obvious, was not an isolated case. If we are to believe the constant parade of statistical evidence issuing from the National Geographic Society and other keepers of the geographical flame, Americans constitute something of a "lost" society.

The most notorious recent example of Americans' collective inability to know where they are and how to get from here to there came out of a Gallup survey commissioned by the National Geographic Society on its hundredth anniversary back in 1988. This survey, which gave people an unmarked map of the world and asked them to locate several selected countries, Central America, and two bodies of water, was designed to test adults in several industrialized nations. Among those surveyed, American adults came in sixth in geographic literacy, with only participants from Italy and Mexico scoring lower than the Americans. The Swedes and West Germans won the gold and silver, with the Japanese taking the bronze in this geographic Olympics. Even more disheartening was the performance of Americans eighteen to twenty-four years old. They finished last among their peers.

What's wrong with our sense of direction? Maybe it is this simple: Americans became geographically stupid when gas stations stopped giving out free road maps.

The *New York Times* columnist Russell Baker once offered another explanation in response to that Geographic Society survey. The problem, according to Baker, stemmed from the availability of R-rated movies and *Playboy* magazine. Before kids could see naked women in movies and magazines, Baker pointed out, they had to turn to the *National Geographic* to get their information. "In the course of the research," wrote Baker, "a good deal of other information rubbed off the page onto the student." And that was before the Internet.

For most people, more likely, it is probably a combination of disinterest and having encountered a Mrs. McNally somewhere along the road in their education. Judging from a sampling of geography textbooks, the materials we provide for learning might contribute to the problem as well. Take this nugget extracted from a geography textbook:

> The internal uniformity of a homogeneous region can be expressed by human (cultural, economic) criteria. A country constitutes such a political region, for within its boundaries certain conditions of nationality, law, government and political traditions prevail. . . . Regions marked by this internal homogeneity are classified as formal regions.
>
> Regions conceptualized as *spatial systems*—such as those centered on an urban core, an activity node, or focus of regional interaction— are identified collectively as functional regions. Thus the formal region might be viewed as static, uniform and immobile; the functional region is seen to be dynamic, structurally active, and continuously shaped by forces that modify it.

Whew! "Internal homogeneity"? "Activity nodes"? "Regional interaction"? If dumbing down textbooks will get rid of this kind of academic gibberish, I say bring on the dumb-downers.

What is so sad about our failure to understand geography is that it reveals a complete misunderstanding of what geography is. In its simplest expression, geography asks humanity's oldest, most fundamental questions:

"Where am I?"

"How do I get there?"

"What is on the other side of the mountain?"

These primal questions have been responsible for pushing humanity from one place to another in search of something better. Eventually, these questions have pushed us off the face of the earth and into the heavens in search of answers to even bigger questions:

"Where do we come from?"

"Is there anybody else out there?"

"Who or what put this universe together?"

Geography doesn't simply begin and end with maps showing the location of all the countries of the world. In fact, such maps don't necessarily tell us much. No—geography poses fascinating questions about who we are and how we got to be that way, and then provides clues to the answers. It is impossible to understand history, international politics, the world economy, religions, philosophy, or "patterns of culture" without taking geography into account.

Geography is a mother lode of sciences. It's the hub of a circle from which other sciences and studies radiate: meteorology and climatology, ecology, geology, oceanography, demographics, cartography, agricultural studies, economics, political science. At some level, all of these can be related to geographic factors. It is obvious that a solid understanding of geography is a vital basic ingredient for a rounded, full understanding of the world and the universe.

Don't Know Much About Geography sets out to ask and answer these questions. This book's simple intent is to make geography a little more interesting than most of us probably recall. The reason so many people don't remember anything about the geography we learned in school is that it was dull. Geography isn't a dusty mystery, but an exciting art as well as a useful science and, like history, it is misunderstood by many Americans. The typical response to these subjects is a glazed eye and an expression like "How dry."

A large part of the problem is that many books about subjects like history and geography are written by experts to be read by other experts. To many of these experts, the common reader and the student are either ignored or approached with utter condescension. Of course, what the experts often overlook in their quest for profundity is the fun in subjects like history and geography.

This book is an attempt to erase typical perceptions of geography. In discussing subjects like history and geography, we get hung up on memorizing FACTS. Dates, battles, speeches, state capitals. Memorizing information is valuable but only if you're able to make some sense of the information and put it into a useful context. Isn't it much better if we can attach something tangible to that information?

All too often, teachers and textbooks forget the human interest in what they teach. Every newspaper editor in the world knows that you must use human interest to sell newspapers. In my previous book, *Don't Know Much About History*, I attempted to make American history a little more appealing and entertaining by emphasizing the personalities and character of historical figures and looking for contemporary references and parallels to give history some connection to our lives. In this book, I have tried to do the same thing by emphasizing the "personality" of geographical concepts and places in the world. Anyone can be taught to memorize where Timbuktu is. But teach people about its location as a junction where the desert and a major river meet and how that led to commercial exchanges that eventually made this ancient city both a learning center and a slave clearinghouse, and then you've made some connection between geography and people's lives.

This book begins with a historical overview of geography that explores the fascinating and frequently amusing subject of human perceptions of the world and the universe through the ages. Just as that group of schoolchildren perceived the Nile River to flow up, the history of the world is littered with other geographical misperceptions and myths that have reflected the attitudes and actions of people throughout history—and often shaped the course of that history.

Of course, the ancients were not the only ones with strange ideas about geography. Maps can be both revealing and misleading. For instance, most people are familiar with a basic world map called the Mercator Projection in which the sizes of continents and countries are vastly out of proportion because a flat map distorts the round earth. In fact, the very depiction of size influences our thinking. While it is the second largest island in the world, Greenland on the Mercator map looks as big as Canada or all of Africa. In fact, it is only 840,000 square miles, compared with Africa's 11 million square miles. Somehow the vastness of the Americas and Africa visually impresses us. But look at them through a population map and they seem "empty" when measured against China or India.

When some event radically alters common perceptions of the world or the universe—Marco Polo returning from Cathay, Magellan's crew completing the circumnavigation of the earth, Lewis and Clark mapping the Louisiana Purchase, the unification of Germany, the disintegration of the Soviet Union or the former Yugoslavia—conventional wisdom is shaken to its foundations and the predictable trajectory of history does a tailspin. Recent events in Europe, as borders are redrawn and countries born or reinvented, provide dramatic evidence of the shaking of geography.

Later chapters explore both the changes in the appearance of the world map brought about by political and historical changes as well as the link between earth's geography and its history.

For sheer pleasure, passages from some of history's memorable travel writers are sprinkled throughout the text as "Geographic Voices." From the ancient Greeks and Marco Polo to astronauts on the moon, travel writers have given the places of the world a vividness that is unmatched by the greatest fiction writers. I have tried to offer an appetizing sample of some of these great travel writers. Also woven through the text are a series of chronologies called "Milestones in Geography" that highlight some of the developments, discoveries, inventions, and events that have shaped the world and people's view of it.

I hope that readers of *Don't Know Much About Geography* will come away from this book a little more comfortable with where things are in the world. But besides being a corrective device meant to refresh the musty memory of everybody's own Mrs. McNally, this book intends to open up the pleasure of geography. The experience is akin to the simple joy a child gets from examining a globe, spinning it with a finger and ending up in exotic places, dreaming far-off dreams.

Beyond that, the book has more ambitious goals. The first of these is to get people to "think geographically," as the ancients did. By that, I mean to see the world with the fullest powers of observation, to look for logical answers, but not to presume that *obvious* and *correct* are the same. To cite a simple example, for many centuries people have looked at the horizon. Many, if not most, assumed the obvious. The earth seemed to come to an end where sky and sea (or land) met. Obvious conclusion:

the earth is flat and if you go too far, you'll fall off the edge. But others looked at that same horizon and made more complex observations: ships going over the horizon seem to sink into it. If the earth were flat, they would simply and gradually disappear. Therefore, the world must be curved. Geographic thinking is another way of saying, look carefully and question the easy assumption.

Thinking geographically also means reading the newspapers with a different eye. Every day there are important events in which *what* happened is directly related to *where* it happened. Anyone who is still stuck with their Mrs. McNally version of events is not going to get the picture.

I also believe that a better sense of geographic literacy might make the world seem a little smaller. Years ago, cable television entrepreneur Ted Turner issued a dictum to writers and newscasters at his Cable News Network—CNN. In Turner's view, the word "foreign" was pejorative and implied peculiar or odd qualities. In the spirit of the "global village," Turner said, the word "foreign" should be replaced by "international" or other alternatives.

A small point of language, perhaps, but a valuable perspective. By making the world a little more familiar, this book aims to help make the rest of the world seem a little less "foreign." That is what *New York Times* columnist Thomas Friedman is pointing to when he talks about the "Golden Arches Theory of Conflict Prevention" in his fascinating book *The World Is Flat: A Brief History of the Twenty-First Century.*

Finally, understanding geography will, I hope, make people understand the tender connections that keep the earth alive. We live in an era in which people truly control the "fate of the earth." For forty years we all worried a great deal about the world ending in a bang. Fears of nuclear holocaust are now lower than at any time since the dawn of the atomic age. But we are faced with threats to the planet that, while not as instantly catastrophic as nuclear war, jeopardize the future of life on earth. Unfortunately, a great many environmentalists have been painted as alarmists with left-field ideas that cost people jobs. In the course of my research, I have become increasingly convinced that a wide range of environmental hazards confronts the future of humanity. And you must understand geography to realize that what happens in the rain forests of Brazil, China's coal country, or the fringes of the Sahara Desert affect life in New York, Kansas City, Dallas, and Seattle. As the phone company liked to tell us in an advertisement of an earlier age, "We're all connected"—and that was in the pre-cell phone era.

Which raises another question. In this age of smartphones, iPads, tablets, cell phones, supersonic travel, simulcast programming that instantly links distant places, cookie-cutter shopping malls that sell the exact same products in Oklahoma City that are sold in Pittsfield, Massachusetts, and McDonald's, Starbucks, and Disney stores sprinkled around the world, does geography even matter anymore? Is geography dead?

It seems a fair question after watching East Berliners, who were allowed to cross into the western sector before the Berlin Wall fell, searching for the Pampers and Burger King Whoppers they had seen advertised on West German television. Or as columnist Friedman more recently wrote, no two countries with McDonald's franchises ever went to war.

But of course, geography will always matter, because for all of the modern world's sophisticated connections, people still prize their individuality, their separateness. The talk of a "new world order" that followed the first Gulf War and the end of Soviet Communism and the difficulties facing the European Union in 2012 is rather empty when people are still killing each other over borders and disputed territory and the world's "have-nots" still look accusingly at the "haves." Technology may shorten distances, but the differences remain. If we ever truly hope to completely bridge those distances and honor those differences, we're going to have to learn the lessons of geography.

DON'T KNOW

MUCH ABOUT

GEOGRAPHY

1

The World Is a Pear

I always read that the world, land and water, was spherical. . . . Now I observed so much divergence, that I began to hold different views about the world and I found it was not round . . . but pear shaped, round except where it has a nipple, for there it is taller, or as if one had a round ball and, on one side, it should be like a woman's breast, and this nipple part is the highest and closest to Heaven.

—Christopher Columbus,
from the log of his third voyage (1498)

Who "Invented" Geography?

Who Made the First Maps?

Imaginary Places: Was There an Atlantis?

Where Was the Garden of Eden?

Who Invented the Compass?

*Why Didn't the Chinese, the Africans, or the Arabs
   "Discover" America?*

Who Did "Discover" America?

Milestones in Geography I: 5000 BC TO AD 1507

In a fleeting instant of historical time, the world has seen transforming events flit across its television screens. The crumbling of the Berlin Wall and, with it, the unification of West and East Germany. The war for Kuwait in the Persian Gulf. SCUD missiles flying into Israel. Arabs and Israelis talking peace. Serbs and Croats at each other's throats. Armenians and Azerbaijanis killing one another over a centuries-old conflict. And most extraordinary of all, the demise of the Soviet Union as we have known it for most of this century.

In the United States, the cover of *Newsweek* magazine asks, "Was Cleopatra Black?" And elsewhere across America, on campuses and in state education departments, debates rage over the multicultural curriculum, emphasizing the historical roots of diverse ethnic groups, and "Afrocentrism," a field of study that emphasizes the contributions of early African civilizations. At the same time, many Americans seek new labels for themselves: African American, Lithuanian American, Ukrainian American.

Suddenly, geography commands center-stage attention, because at their heart, all of these issues are questions of geography.

During his 1988 campaign for the White House, George H. W. Bush often said he wanted to be known as the "education president." But as one Washington wit said in the midst of the Gulf War in 1991, "We didn't realize he was going to teach us geography."

If nothing else, the world certainly did get a thorough geography lesson during Desert Storm, as nightly newscasts and special bulletins from the front lines in Kuwait showed detailed maps of the Middle East and the Persian Gulf. With a daily diet of military press briefings by a host of generals and air marshals, names and places once familiar only from a distant past of childhood fairy tales or Sunday school Bible lessons suddenly became household words: Baghdad, Arabia, Jerusalem.

The global village never seemed so small. And people around the world who never gave much thought to maps—except when they needed to find their way to a vacation spot or to puzzle out the mysteries of the New York City subway system—were looking at world maps with new eyes, even as those maps were being redrawn.

All at once, Americans, along with the rest of the world, were contemplating geography, perhaps for the first time since leaving elementary school. Unfortunately, for most of those people, the word *geography* conjures up images of musty textbooks, or being forced to memorize the names of capitals, or elementary school assignments in which you pasted a cotton ball on maps of Alabama and Mississippi, a copper penny on Utah, and a grain of rice on China to show the chief products of these locations.

But now geography—or thinking "geographically"—has been thrust upon us. We can no longer afford the blissful ignorance of thinking of the world in the terms of the famous *New Yorker* poster by artist Saul Steinberg in which New York fills the foreground while the rest of America and the world beyond appear as insignificant bumps on the horizon.

The irony of this modern inability to think geographically—or sheer disinterest—is that it is so far removed from the thinking of the past. From the earliest moments of human history, people have had to think geographically in order to survive and for the world to progress as it has. It was that ability to observe the world and make reasoned conclusions about the earth and the universe itself that began the march of science.

Geographic Voices　Aristotle (384–322 BC)

> Furthermore, the sphericity of the Earth is proved by the evidence of our senses, for otherwise lunar eclipses would not take such forms; for whereas in the monthly phases of the moon the segments are of all sorts—straight, gibbous, and crescent— in eclipses, the dividing line is always rounded. Consequently, if the eclipse is due to the interposition of the Earth, the rounded line results from its spherical shape.

Who "Invented" Geography?

Imagine this. You've been shipwrecked and washed up on a desert island, a modern-day Robinson Crusoe. A selective amnesia has erased any memory of dates, places, seasons, or time. You have no watch, no maps, and no recollection of where you were when your ship went down.

How long would it take you to figure out the time of day? The season? The month? The approximate date? You notice that the water comes way up onto your beach and then goes back out later in the day. Why does it do that? As you lay back in your tropical paradise and looked at the night sky, could you distinguish among those pinpoints of light that moved through the heavens?

When would you plant some crops to keep yourself fed? After all, coconut milk and wild berries only go so far.

Do you know the distance to the other side of the island? How would you measure it? And what about your approximate location in the world? You've forgotten latitude and longitude exist. Do you know where in the world you are?

If you managed to figure out all that, could you then determine what shape the world is? And how large that world might be?

Well, the ancient Greeks—or more accurately, a varied group of people we have lumped together and called the Greeks—managed to do just about all of these things. Of course, it took several geniuses working over the course of a few centuries to pull all of this together— and not without a few substantial mistakes that were kept alive for most of the next twenty centuries, influencing everyone from the hierarchy of the Roman Catholic Church to Christopher Columbus.

But the Greeks did it. And they managed it without watches, telescopes, sextants, Black & Decker tape measures, or any of the other useful little devices that have made accurate measurement of time and space possible. The Greeks were not the first people to look at the world and attempt to explain its workings. The Egyptians and Mesopotamians produced much of the groundwork from which the Greeks proceeded. And the Indian and Chinese cultures were working things out in their own way for much of the same time.

But what set the early Greek thinkers apart from their contemporaries as well as from earlier cultures was their systematic attempt to apply rational thought to the world. They were the first to explore the notion of testing their ideas about the world in the beginning of what we now call the scientific method. And while they fell back on myth and superstition when they were unable to explain the universe—just as past and future generations of humanity would—they were the first to attempt to *know* the universe.

Geography is a word derived from the Greek—*ge*, meaning "the earth," and *graphe*, "to describe." Many Greeks thought and talked and wrote about geography without exactly calling it that. In fact, Homer's epic *Odyssey* is viewed as one of the first geographic works in Western culture because it describes the many recognizable places that Odysseus (Ulysses) visited during his long voyage home from Troy. (See Chapter 4, page 166, "Imaginary Places: Was There a Troy?")

More scientific approaches to geography came about in Miletus, a Greek trading center that flourished some seven hundred years before Christ in what is now modern Turkey. There, Greek philosopher-mathematicians began to apply mathematical principles to measuring the earth. Thales, a sort of ancient Thomas Edison, combined his success in the olive oil business with an extraordinary ability to both ponder and invent. He made several major contributions to geometry and was said to have accurately predicted a solar eclipse in 585 BC. But one of his influential conclusions was that the earth was a disk floating in water.

Anaximander, a younger colleague who introduced a sundial, made a rather astonishing guess when he surmised from fossil remains that life originated in a sea that once covered much of the earth's surface. He drew the first scaled world map. With Greece in the center, it showed a world bounded by an endless river or sea. He believed that the earth was a cylinder with a disk, the habitable part, resting on top. But instead of floating on an endless sea, as his mentor Thales had thought, Anaximander's earth was suspended freely in space; the heavens were attached to a sphere that revolved around the earth, which explained the daily circuit of sun, stars, and planets.

Other Greek writers, philosophers, historians, and mathematicians followed—Herodotus, Plato, and Aristotle among them—all expanding the Greek inquiry into the size and shape of the world, its place in the universe, and the bounds of human habitation. Plato believed the earth was spherical, but for philosophical reasons, not through scientific evidence; the sphere, he believed, was the perfect geometric form. Aristotle later agreed, but sought observable evidence, which he found in the shadow cast by the earth on the face of the moon.

On the other hand, the great philosopher also fell back on fairly simplistic reasoning. Aristotle thought that the closer to the equator, the hotter the temperature. His "proof" lay in the black skin of Libyans who, in Aristotle's thinking, had been seared by the sun. Life at the equator was not possible, in Aristotle's conception, because it would be too hot there. Aristotle also believed in a natural balance that dictated the existence of a continent to the south of the equator equal to one north of the equator, introducing the concept of the antipodes, or "opposite feet," that lasted from Aristotle's time until the voyages of Captain Cook in the mid-eighteenth century.

But three other so-called Greeks stand out because they all addressed Greek knowledge of the world in separate books, all with "geography" in the title.

The first of these was Eratosthenes (circa 276–194 BC), actually a Libyan-born librarian who was the first to use the word "geography" and who also managed to come up with a way to measure quite accurately the circumference of the earth with little more than a shadow, a well, and some basic camel sense.

Eratosthenes was appointed chief of the library at Alexandria, where he controlled a collection of more than a hundred thousand "books"—actually papyrus scrolls—containing the known world's collective knowledge. About two hundred fifty years before the birth of Christ, the Western world's most important city was Alexandria, in Egypt, the home of the renowned library started by Alexander the Great, the young soldier from Macedon tutored by Aristotle. After Alexander's death, his heirs as rulers of Egypt were the Ptolemies (the legendary Cleopatra among them). Under the three-hundred-year-long Ptolemy dynasty, Alexandria became the world's preeminent center of scientific, mathematical, and literary studies, as well as a rather seamy den of cutthroats drawn by the riches of the world that passed through the city. It was, as one poet called it, the "house of Aphrodite" (goddess of love) with plenty of wine, wealth, fine young men, and beautiful women. Makes you wonder how they got any work done.

One of Eratosthenes's greatest contributions seems simple enough, given the benefit of hindsight. However, nobody else thought of it sooner, so Eratosthenes gets the credit for dividing the world by parallel east-west and north-south lines, or meridians. He failed to lay these lines down at regular intervals and instead used notable landmarks and prominent places such as Rhodes, Alexandria, the Pillars of Hercules (Gibraltar), and the tip of the Indian Peninsula as the basis for dividing his world.

Hearing of a well in Syene (modern Aswan) where the sun's reflection could be seen in the water at noon on June 21, the longest day of the year, Eratosthenes surmised that the sun was directly above the earth at that moment. The Libyan librarian then made some interesting logical leaps. He believed that Syene was due south of Alexandria on the same meridian (or longitudinal line, the imaginary north-south running lines on the map). By measuring the shadow cast by an obelisk in Alexandria at the same moment there was no shadow in Syene, Eratosthenes computed the length of two sides of a triangle—the length of the shadow and the height of the obelisk. With that information and some basic geometry, Eratosthenes figured the angle of the triangle, and with that figure determined the degree that the sun was from directly overhead. That proved to be 7°12, which is approximately equal to one fiftieth of a circle's 360 degrees.

Knowing this, Eratosthenes further reasoned that if he knew the distance from Syene to Alexandria—which would equal the third side of his triangle connecting the sun, Alexandria, and Syene—he could simply multiply that distance by fifty to get the approximate size of the earth. Enter the camels.

Eratosthenes learned that it took a camel caravan fifty days to make the trip from Syene to Alexandria. Using ancient EPA camel standards of 100 stadia per day (stadia is an ancient measurement that related to the size of a Greek race course), the clever librarian came up with a distance of 5,000 stadia from Syene to Alexandria. Multiplying that by 50 gave Eratosthenes an earth circumference of 250,000 stadia. Using various estimates of modern equivalents, his earth measured about 25,000 miles—very near its actual measurement at the poles of 24,860 miles. Given the number of small mistakes involved, all of which canceled each other out, this calculation was an extraordinary example of the Greek ability to apply logic and mathematics to measuring—and knowing—the world.

After Eratosthenes died, a conflicting view of the size of the earth came from another Greek historian-geographer named Posidonius (circa 135–51 BC) of Rhodes. His calculation was based upon the height of the star Canopus, determined algebraically, and used the sailing time of ships. Ironically, his calculations were close to the figure Eratosthenes had reached. But for some reason they were later reduced to the much smaller size of 18,000 miles by Strabo, another significant scholar who comes along next. It was this figure that Columbus would rely upon in making his case for a voyage west to the Orient.

While this mistake—a smaller earth—was widely accepted and perpetuated, another of the conclusions reached by Posidonius was correct, but dismissed because it contradicted Aristotle. Posidonius believed that the equatorial zone was quite habitable and that the highest temperatures were to be found in deserts inside the so-called temperate zone, which is the case.

The expert who inaccurately recorded Posidonius was the second key "Greek" geographer, Strabo (circa 64 BC–circa AD 20), who was born in what is now modern Turkey and who wrote at about the time of Christ. Like Eratosthenes, Strabo worked in the Alexandria library. Unlike Eratosthenes, Strabo was no innovative genius who came up with new theories about the world. His genius, instead, was as a compiler and his work, *Geographica*, filling all of seventeen volumes, brought together the sum of the Mediterranean world's knowledge to that time, describing Asia, North Africa, and much of Europe, which Strabo had seen himself in his extensive travels. Among his chief contributions was his division of the world into frigid, temperate, and tropic zones, although he badly miscalculated how far north and south of the equator these lands were habitable. He believed, like Aristotle, that the dark skin of the Ethiopians was the result of scorching by the sun and that the blond barbarians of the north were savage because of the frigidity of the arctic zones.

And finally, there was Ptolemy (circa AD 100–170), an Egyptian-Greek or Greek-Egyptian (but not one of the royal Ptolemies) who condensed the sum of Greek world knowledge during the period of the Roman Empire and whose views were accepted for centuries.

Although best known for his work in the area of astronomy—kept alive by the Arabs and known by its Arabic name, *almagest*—Ptolemy's *Geographia* laid out many of the principles still followed in modern cartography and included an atlas of the known world, based on the reports of the Roman legions as they spread the Roman Empire. It included some eight thousand places identified by their latitude and longitude—words Ptolemy is said to have coined. And the system he adopted is basically that of modern geography, including the seemingly simple notion of orienting maps with the north at the top and the east on the right. That is, it seems simple enough until you realize that if you set out to orient a map today, what would you place at the top? Given the notion that the world is a sphere, any arbitrary spot might have been used. For many centuries, for instance, European maps were oriented with east on top, emphasizing the centrality of the Holy Lands, and Jerusalem in particular.

Ptolemy also attempted to address a problem that still exists: the impossibility of representing a round earth on a flat piece of paper. His solution was a globe, but that posed its own problems, as a globe could not be made large enough to encompass the fine details that Ptolemy wanted to include in his maps.

Ptolemy's world was surprisingly large, consisting of the three continents then known to the people around the Mediterranean—Europe, Asia, and Africa. Although often inaccurate in matters of size, shape, and precise location, it included the British Isles, Scandia (Scandinavia), and Sinae (China). He also described the source of the Nile quite accurately as lakes in Africa south of the equator—hidden in the "Mountains of the Moon"—a fact left unproven to the European world until the travels of the British explorers Sir Richard Francis Burton and John Speke in the nineteenth century.

Like his predecessors, Ptolemy made mistakes, and these influenced the course of science, philosophy, and religion. His earth-centered universe would be accepted by the learned world for centuries to come. He elaborated on the concept of the antipodes, expanding it into a Terra Australis Incognita (unknown southern lands) which, on one hand, fueled speculation and hope of finding a great continent attached to the bottom of Africa but, on the other hand, made sailing around Africa seem impossible.

But one of his mistakes was even more far-reaching. Relying upon Strabo's figures, Ptolemy declared the world to be eighteen thousand miles around. On his maps, Asia extended far beyond its true width, making the Orient seem far closer to Europe than it actually is. Ptolemy's authority, like Aristotle's, was unquestioned by later Europeans, including most significantly one Genoan named Cristóbal Colón, who used Ptolemy's figures to argue his case before the king and queen of Spain.

Geographic Voices   From Strabo's *Geographica*, written between AD 17 and 23.

The Amazons are said to live among the mountains above Albania. . . . But other writers say that the Amazons bordered upon the Gargarenses on the north, at the foot of the Caucasian mountains. . . .

When at home they are occupied in performing with their own hands the work of plowing, planting and pasturing cattle, and particularly in training horses. The strongest among them spend much of their time in hunting on horseback, and practice warlike exercises. All of them from infancy have the right breast seared, in order that they may use the arm with ease for all manner of purposes, and particularly for throwing the javelin. . . . They pass two months of the spring on a neighboring mountain, which is the border between them and the Gargarenses. The latter also ascend the mountain according to some ancient custom for the purpose of performing common sacrifices, and of having intercourse with the women with a view to offspring, in secret and in darkness, the man with the first woman he meets. When the women are pregnant they are sent away. The female children that may be born are retained by the Amazons themselves. . . . Where they are at present few writers undertake to point out.

Who Made the First Maps?

Even those Greeks, as extraordinary as they were, had help. Long before the rise of the classical Greek period around 500 BC, plenty of other people were observing the heavens and the earth, then drawing some amazing conclusions about the workings of the universe. The Greeks, a trading and seafaring people who came into contact with other civilizations, were quick studies.

One of those groups encountered by the Greeks was very good at figuring out the heavens. They also left some of the earliest maps as well as the first "world map." These were the people of the so-called cradle of civilization, the ancient inhabitants of Mesopotamia whose descendants made the newspapers in 1990 and 1991 in explosive fashion in the country we now call Iraq.

Set between the Rivers Tigris and Euphrates (*Mesopotamia* is Greek for "the land between two rivers") lay a broad, fertile valley. Though much of it is desert today, ten thousand years ago it was covered by ice-age grasslands. It was here that nomadic tribes of hunters followed herds of grazing animals. As the ice caps retreated, deserts replaced the grazing land, and the nomads were drawn to the valley where the annual flooding of the two rivers provided water and food for the animals. Over thousands of years, these people learned the secret of sowing cereals in the mud beside the rivers—the birth of agriculture.

By 8000 BC, the people of Mesopotamia—first the Sumerians and in later centuries the Akkadians, Babylonians, and Assyrians—were using clay tokens to record numbers of animals and measures of grain, the rudimentary beginnings of written numbers and language that would develop over the next five thousand years. Eventually, they produced such significant innovations as the potter's wheel and wheeled vehicles, kiln-fired bricks, bronze smelting, and beer.

The first known "map of the world" is a Babylonian clay tablet that dates from approximately six hundred years before Christ. This flat disk is tiny—about three by five inches—and depicts the world as a circle with two lines running down the center, representing the Tigris and the Euphrates Rivers. Encircling this is the Bitter River. Outside its bounds reside imaginary beasts, a mapmaker's work of the imagination to signify the unknown, a tradition that continued for many centuries to come.

While that clay tablet from ancient Babylon represents what has been called the first known world map, there are much earlier maps from this area and others. A map of the Mesopotamian city of Lagash is carved in stone in the lap of a statue of a god, the oldest known "city map." Clay tablets showing settlements and geographic landmarks have been found in northern Iraq and dated to 2300 BC, the period of Sargon I of Akkad. These maps, and others from about the same time in Egypt, show plans that undoubtedly were used for assessing property taxes! This seems to confirm the old cliché about the only certainties in life.

More recently, a relic that might be called a map was discovered in Mezhirich in the former Soviet Union. This piece of ancient bone on which etchings have been made is estimated to come from a time ten to twenty thousand years before Christ. It is presumably an early road map of sorts, showing the region around the site at which it was discovered. It is safe to assume that rudimentary maps predate written language, as the earliest humans scratched out symbols to show their neighbors the way to the happiest hunting grounds.

But the sophistication of the early Babylonian maps are testimony to the advances made by the people between the Tigris and the Euphrates Rivers. And while their maps stand as evidence of their sophistication, their true brilliance was in their study of the heavens, marking the beginnings of astronomy.

It seems ironic that much of what these early people knew and understood about the earth derived from their understanding of the heavens, which is why astronomy has often been called the "first science." Nowadays people, especially those urbanites unaccustomed to open spaces, marvel at a night sky filled with stars. But because of pollution and the brightness of man-made illumination, there are few places left in the world where the brilliance of the night sky equals the celestial canopy witnessed by the ancient people who began to observe the motion of the sun, stars, and moon and began fixing their seasons to these regular movements.

Another group that excelled in mapmaking was the Chinese. From well before the time of Christ until some fifteen hundred years later, the Chinese enjoyed the world's highest standard of living. Prosperous and agriculturally rich, China was a well-organized, comfortable society, far advanced in science and practical invention. Chinese mathematicians may have developed the zero and the decimal system and introduced it to the Hindus, who passed it to Baghdad. Surpassing the Mesopotamians, Chinese astronomers kept the longest and most continuous records of celestial events, and Chinese records mention such events as the appearance of a comet in 2296 BC and the explosion of a supernova (the rare explosion of most of the material in a star, resulting in an extremely bright, short-lived object emitting vast amounts of energy) in 352 BC.

They also elevated cartography to a beautiful art, as well as a far more developed science. Although there are references to maps in Chinese literature dating from 700 BC, the oldest Chinese maps yet found date from 200 BC. Highly accurate and incredibly detailed, these maps, woven of silk, showed the names of provinces, used symbols to distinguish between towns and villages, and depicted mountain ranges and the courses of rivers and roads. Another early Chinese map—again testifying to their advancement in cartography—described the military defenses of the kingdom in impressive detail.

The common denominator among these early maps is the prevalent sense that the mapmaker existed in the center of the universe, an attitude still evident in many people, but particularly those who live in Manhattan or Paris!

By looking at the maps of the world or the myths they often represented from these cultures, it is apparent that people have always carried an inflated sense of our place in the universe. In almost every society, from the ancient Babylonians, Egyptians, Greeks, and Chinese to the Aztecs and Plains Indians of the Americas, humans have placed themselves at the center of the world. Beyond that, until fairly recently in human history, people went a step further and put the earth in the center of the universe, with sun and stars circling this rather insignificant speck in the vast cosmos.

This tremendous sense of self-importance that led the world of rational people to hold on to the idea that the earth was the center around which the rest of the universe spun, is a common human foible that has been labeled the *omphalos syndrome*, from the Greek word for navel. The Greeks placed their omphalos at the famed Temple of Delphi located on the lower slopes of Mount Parnassus. Here was the temple seat of the most important oracle in Greece. Considered to be the center of the world, the Delphic oracle was consulted on all matters of state. Similarly, the word Babylon comes from an ancient word meaning "door of the gods," or the spot at which the gods came to earth.

Of course, when you believe that you exist at the center of the world and don't know much about the rest of the world, the realm of observation and logical speculation blend into the world of mysticism with results that are at least amusing to the twentieth century. The worldviews of these ancient cultures reflected their geographic sense as much as it did their philosophies. In fact, the two were closely connected. For the Egyptians, the Nile dominated their lives and their view of the universe. The Nile and its regular flooding were life itself. The Egyptian world was divided in two by the Nile, which flowed into a great ocean. The sky was held up by four supports; sometimes depicted as poles, sometimes mountains. Beliefs about the sun's daily course took various forms from everyday life: a hawk rising each day, or the sun pushed along by a giant beetle, just as a beetle rolls its ball of dung. In another version, the sun god Ra drove his chariot—adopted later by the Greeks as Apollo.

The Babylonians, whose lives were also dominated by their two rivers, believed there was an immense body of water within the earth that gave life to the world. With its emphasis on astronomy, the Babylonian view of the universe envisioned a great vault of the heavens over the earth. This idea found expression in Chinese views as well.

Often, as with Mount Olympus, the home of the Greek gods and usually hidden from view, the center was a mountain. For the Japanese, it was Mount Fuji. In Hindu and later Buddhist cosmology, influenced by its proximity to the seemingly unscalable Himalayas, a mythical Mount Meru rose up as the center of the earth. The home of the gods, this mountain rose eighty-four thousand miles into the heavens.

IMAGINARY PLACES: Was There an Atlantis?

All of these ancient notions of the world point up the dividing line between scientific, rational processes and the leap into legend, imagination, or faith. Even the Greeks, for all of their incredibly developed notions of the world, fell back on myth, superstition, and legend to explain those things for which there was no verifiable and rational explanation.

One of the most familiar examples of this from the Greek era is a geographic myth that persisted for centuries, embroidered upon through history until it became a part of modern consciousness, even contributing to the success of a rather mindless pop song of the sixties by a singer named Donovan. That legend was the story of the mythical Atlantis, and the source of this long-held myth, as unlikely as it may seem, was Plato.

Most of what the world has thought about the existence of Atlantis as a superior culture that disappeared during a sudden cataclysm comes from two of Plato's dialogues, *Timaeus* and *Critias*. Plato said he got the story from Socrates, who heard it from Solon, who was told it by the Egyptians. You can see how the story might have changed as it went along.

According to Plato's account, Atlantis once occupied a large island west of the Strait of Gibraltar (or Pillars of Hercules, to the Greeks). In this legendary civilization, which Plato claimed had flourished nine thousand years earlier, men descended from the sea god Poseidon had created an earthly paradise. Food was plentiful, the buildings and temples were magnificent in architecture and embellishment. A temple, for instance, was "coated with silver save only the pinnacles and these were coated with gold. As to the exterior, they made the roof all of ivory in appearance, variegated with gold and silver. . . ." As Plato described it, Atlantis was a great military power that could muster an army of one million and was preparing to assault Athens and Egypt when the great disaster struck.

In Plato's version:

> At a later time, there occurred portentous Earthquakes and floods, and one grievous day and night befell them, when the whole body of your warriors was swallowed up by the Earth and the island of Atlantis in like manner was swallowed up by the sea and vanished; wherefore also the ocean at that spot has now become impassable and unsearchable, being blocked up by the shoal mud which the island created as it settled down.

Over the centuries, the legend of Atlantis grew. While in Plato's account it was destroyed suddenly, the mythical Atlantis lived on, prospering as life on the island miraculously continued under the sea. Even the modern mystic Edgar Cayce spoke of the disappeared continent and prophesied the imminent reemergence of Atlantis!

In fact, the legend of the lost continent of Atlantis is probably based on the fate of the island of Thera (also called Santorini), about seventy miles north of Crete in the Aegean Sea. From geological and archeological records, it is known that sometime between 1650 and 1500 BC, the volcano Santorini erupted and destroyed most of this island, leaving a small rim of rock on one side of a large water-filled caldera (a large basin-shaped crater, usually formed when a volcano subsides and all but the top is covered by water. In North America such a caldera stands as a small island in Crater Lake, Oregon).

The eruption and ensuing tidal waves have been blamed for the eventual end of the civilization of the Minoans, a people who lived on Crete at about this time and were named for their legendary King Minos. The Minoans have been recognized as one of the richest, most powerful, and most advanced peoples of the ancient world, a description that matches up nicely with Plato's account of the dwellers of Atlantis. It was the Minoans who had elaborate religious ceremonies involving bull worship that gave rise to the familiar myth of the Minotaur and the Labyrinth. After 1400 BC, the Minoan civilization vanished from the records. One popular theory holds that Santorini's eruption signaled the onset of the demise of Minoan civilization.

Because of its likely date, the eruption of Santorini has also been suggested as a possible natural cause of the biblical plagues on Egypt recounted in Exodus and the subsequent parting of the Reed Sea—not the Red Sea, as we were taught for so long—by Moses. Although this theory is more controversial, it is an intriguing one, with the volcanic ash and tidal waves spawned by the Thera eruption accounting for some of the natural phenomena described in Exodus. The difficulty with this suggestion is in dating the Exodus. Usually it is placed in the thirteenth century BC, rather than the sixteenth century BC, during which time the volcano most likely erupted.

Where Was the Garden of Eden?

If the Atlantis myth persisted for centuries, fueling speculation and superstition, it was only typical of the human fondness for mixing reason and faith. All of the geographic myths of the ancient cultures were tied to the entire philosophical or religious system of these societies and are best illustrated in their creation myths. Every society has a creation story because all people want an explanation for their beginnings. Usually those beginnings are tied to some special status for the group. It is easy to proclaim your superiority when you can tell people you are a product of divine intervention.

The creation myth with the greatest impact on Western civilization has been the biblical story of the Garden of Eden.

Then the Lord God formed man of dust from the ground, and breathed into his nostrils the breath of life; and man became a living being. And the Lord God planted a garden in Eden, in the East; and there he put the man he had formed. And out of the ground the Lord God made to grow every tree that is pleasant to the sight and good for food, the tree of life also in the midst of the garden, and the tree of knowledge of good and evil.

A river flowed out of Eden to water the garden, and there it divided and became four rivers. The name of the first is Pishon; it is the one which flows around the whole land of Hav'ilah, where there is gold; and the gold of that land is good; bdellium and onyx stone are there. The name of the second river is Gihon; it is the one which flows around the whole land of Cush. And the name of the third river is Tigris, which flows east of Assyria. And the fourth river is the Euphra'tes.

The biblical version of creation found in Genesis, with its earthly paradise, the Garden of Eden, from which Adam and Eve are eventually expelled ("He drove out the man; and at the east end of the garden he placed the cherubim, and a flaming sword which turned every way, to guard the way to the tree of life."), was the source of inquiry, speculation, and searching for much of early Christian-era history.

While modern Christians debate the degree to which the biblical version of history should be literally accepted, for European Christians of the Middle Ages there was no doubt. Scripture was simply the divinely inspired word of God, about which there could be no legitimate question. To treat these accounts with even the slightest uncertainty was heresy, frequently a deadly career choice.

With that in mind, geographic studies in medieval Europe moved away from the Greek tradition of expanding scientific and geographic knowledge to an age of faith obsession with explaining the world through biblical lessons. Unfortunately, some of the inspiration for these holy-minded men came from less-than-scriptural sources. One reason was the existence of a far more appealing version of world geography, produced by such influential writers as Pliny the Elder (AD 23–79), Lucian of Samosata (circa AD 120–190), and Gaius Julius Solinus (active circa AD 250).

Pliny's *Natural History* was of very dubious scientific merit, but his ideas about the world were a mainstay of mapmakers for centuries to come. Although a dry compiler of known places, Pliny became downright enthralling when he got to describing the world beyond his own experience and firsthand knowledge. Among the marvelous places and people he said existed were the people of the Ear Islands, which were situated off the coast of Germany, a tribe of fishermen whose ears were so large that they covered their bodies. These large appendages offered the benefit of enabling the All-ears to hear fish under the sea. On Evileye Island, near the North Pole, the women were possessed of a stare that could bewitch or even kill. Pliny warned travelers of Hyperborea, an island with cliffs shaped like women, which came to life at night to destroy ships. In this land near Scotland, the sun rises and sets only once a year. Although sorrow is unknown in Hyperborea, people chose the time of their own death and jumped into the sea from the Leaping Rock. On Lixus, an island off Africa, Pliny told of a tree that bore golden fruit. On the isle of Taprobane, there were snakes with a head at each end of their bodies. And in the desert of Africa, he told of the Blemmyae, a race of headless people whose eyes and mouth were located in their chest.

Unlike Pliny, Lucian did not take himself seriously. His satirical works marked the beginning of a tradition that later attracted such writers as Voltaire, Rabelais, and Swift; writers who used the traveler's tale to skewer contemporary habits. Despite its obvious fabrications, Lucian's *True History* eventually found its true believers. Among the wonders he told of were Caseosa, or Milk Island, a twenty-five-mile round isle of cheese where the grapes produced not wine but milk. On Dionysus's Island there were vines shaped like women that could speak. However, travelers were advised not to converse with these vine-women because the men would become drunk. An unlucky soul who attempted sexual intercourse with one of these creatures risked transformation into a vine himself. On the Atlantic isle of Cork lived the Corkfoots, who, as their name suggested, could walk on water with their feet of cork. Pumpkin Island featured pirates who sailed out on boats carved from huge pumpkins.

While Lucian was a bit of an ancient Merry Prankster, others who followed took such tales with utter faith. The attempt to reconcile science with Holy Writ, and rationalism with religion, began haltingly with Solinus, who lived around AD 250 and without apology cribbed his material straight from Pliny. He told of horse-footed men, one-eyed hunters who drank from cups made of skulls, and the umbrella men of India whose one leg ended in a foot that was large enough to cover their heads. There were also interesting animals in the world of Solinus, none more so than the lynx whose urine turned into a precious stone with magnetic powers. One of the real contributions of Solinus was his decision to rename the familiar waters around Rome. Long called Mare Nostrum (Our Sea), Solinus introduced his preference for Mediterranean, or "Sea in the Middle of the Earth," once again reflecting the omphalos syndrome. Was there any question for a Roman citizen at the peak of the empire that he lived in the center of the world?

Perhaps not surprisingly, these geographic tall tales were widely accepted as the European world was plunging into the medieval period and the emphasis among scholars shifted from science to faith. One who was greatly influenced by what Solinus had written was St. Augustine (AD 354–430), bishop of Hippo in the Roman province of Algeria and the most influential Christian thinker for the next several hundred years. Familiar with the literature of the pagan (Greek) world, Augustine struggled with the contradictions between Scripture and the classics. The Bible names only three continents—one for each of Noah's descendants; how could there be another? On the question of the antipodes, Augustine was decisive; there was no rational ground for such a belief.

As time went by, Ptolemy's reasoned ideas of a sphere with well-marked grids showing latitude and longitude were forgotten. In their place, the extraordinary worlds of Pliny and Solinus took hold. The Greek sphere was replaced by a new Christian vision in which the world, in one famous rendition, was rectangular, like a treasure chest with a vaulted top to hold the heavens. The author of this particular version of the world was Cosmas Indicopleustes ("India traveler"), a widely traveled merchant turned Christian mystic of the sixth century. To Cosmas, all matters of the size and shape of the universe, as well as descriptions of its inhabitants on earth, could be found in a careful reading of the Bible. For him, such an idea as the antipodes was simply ridiculous. He asked, "Can anyone be so foolish as to believe that there are men whose feet are higher than their heads, or places where things may be hanging downwards, trees going backwards, or rain falling upwards?"

Cosmas's belief in Eden was firm, unshakable. However, it lay beyond the ocean, unreachable by men.

Even more influential than Cosmas was Isidore of Seville, who lived in the depths of the Dark Ages in the sixth and seventh centuries. Isidore was more precise in his notions of the location of paradise. Author of an influential encyclopedia of the knowledge of his time, Isidore placed paradise firmly on an island in far eastern Asia. Its location would be included on almost all maps in the Western world for hundreds of years to come. According to Isidore, there was only one problem in reaching paradise. God, as Genesis clearly stated, had closed off all approach to Eden with a swordlike flame.

The search for Eden powered myths and legends that stayed alive for centuries. One of these was the story of St. Brendan (circa 484–578), a sixth-century Irish monk who supposedly "discovered" America nine hundred years before Columbus sailed. Inspired by a dream, Brendan set out to find paradise. With a crew of sixty, he spent five years at sea, encountering strange beasts, including birds who told the saint that they were fallen angels. And on a lonely rock in the midst of the ocean, Brendan found the solitary figure of Judas Iscariot, the betrayer of Christ. Finally, Brendan reached a beautiful island where he encountered a holy man and a dead giant who came to life. This story, told as a combination of medieval legend and fact, was kept alive for hundreds of years during which time Brendan's "Promised Land of the Saints" was clearly marked on maps, including some that placed Brendan's paradise in the vicinity of where North America was later discovered by European adventurers.

If Eden was truly in the East, as most men believed, the area held another place far more real and perhaps even more meaningful for the medieval European Christian. In declaring the First Crusade in 1095 to recapture the Holy Lands from the grasp of Islam, Pope Urban II reportedly said, "Jerusalem is the navel of the world, a land which is more fruitful than any other, a land which is like another paradise of delights."

The First Crusade, one in a series of wars launched to retake Jerusalem from the Muslims, succeeded in capturing Jerusalem in July 1099. But that was pretty much the end of the crusaders' success. A short-lived one it was at that, as Jerusalem was retaken by Saladin in 1187 (See Chapter 4, "World Battlefields That Shaped History," p. 184.) But the hopes of the European Christians were bolstered by word of a powerful ally against the Islamic "hordes." Around the time of the First Crusade, reports began to circulate of a great Christian king in the East called Presbyter or "Prester" John. Supposedly descended from the three kings of the Bethlehem story, Prester John had supposedly defeated the Persians in an epic battle and was heading west to aid the crusaders. A few years later, a letter from this extraordinary eastern general-king came forth. A complete fabrication, it nonetheless fed hopes that Prester John, like an avenging angel, would come galloping of the East to join forces with the European Christians in their quest to recapture the Holy Land. The story of Prester John, like that of St. Brendan and other medieval myths, had a great shelf life. Two hundred years later, Marco Polo even claimed that Prester John had once existed but was killed in a great battle with Genghis Khan.

Despite its hopelessness, all this searching about for Prester John and warring to take Jerusalem had its practical effects. Hunkered down in isolation since the fall of Rome and cowering under the impact of the plague that struck Europe in the mid-fourteenth century, Europeans were finally getting out and seeing the world. Many of their ideas remained stuck in a murky combination of faith and magic, but the door had opened and allowed a crack of light from the East to shine through.

And the European quest to locate Eden remained real, right up to the time of Christopher Columbus, who was convinced on his third voyage in 1498 that he had found paradise. His description of the earth as pear-shaped came at a time when he reached the mouth of the Orinoco River (in modern Venezuela). Columbus, possibly the son of a Jewish father, was an intensely pious man who took scripture seriously. He believed that this rush of fresh water came from the four biblical rivers of paradise, and that the expanse of land he saw was the earthly location of Eden. He added to his log, "I am convinced that it is the spot of the earthly paradise whither no one can go but by God's permission."

Who Invented the Compass?

One reason Columbus was able to get as far as he did was his possession of what we think of today as a fairly simple piece of technology. Every Boy Scout has one. There is probably one, in some form or other, around every household. Some people keep one on the dashboard of their car. These devices are a source of endless fascination, but most of us don't know what to do with them. They are magnetic compasses. They work because of the earth's strange magnetic poles. These poles, a characteristic feature of this planet, are not exactly the same as the geographic poles. The reason for the location of the magnetic poles remains a mystery, and the earth's magnetic field has apparently reversed polarity many times in the geologic past. Nonetheless, a magnetic needle will come to rest pointing approximately northward. With a card fixed underneath indicating the point of direction, the compass user simply fixes the needle over the north point on the compass card and can determine the precise direction in which to head.

People take these compasses for granted, and like a lot of other small conveniences in life, we don't really understand how they work. In the case of the compass, like the wheel and a hundred other very basic inventions that made human progress possible, we also don't know who invented it. Strange as it seems, no one knows who discovered the magnetic property of the lodestone, or magnetite. We don't know who figured out that the stone's attractive power could be passed on to a piece of iron, or who discovered that the magnet could be used in determining geographic directions.

Like many unexplained phenomena of nature, the lodestone was the source of legends before its practicality was discovered. One such legend was the existence of magnetic islands through which ships could not pass if they were built with nails. From earliest times, the remarkable powers of the lodestone—the word comes from an obscure Old English word for *way*, so a rough translation is "stone that shows the way"—were associated with dark, magical forces.

In medieval Europe, magnets were in fact used by magicians to perform crowd-pleasing tricks. According to another legend, a piece of lodestone placed under the pillow of an unfaithful wife could make her confess her sins. The mineral was said to be so potent that a small piece could cure all sorts of ailments and even act as a contraceptive. According to Simon Berthon in his book about map history, *The Shape of the World*, the compass was also said to "have the power to reconcile husbands to their wives, and recall brides to their husbands."

And in *The Discoverers*, Daniel Boorstin recounts, "Since this inexplicable power of a magnetized needle to 'find' the north smacked of black magic, common seamen were wary of its powers. For many decades, the prudent sea captain consulted his compass secretly. . . . After the compass had lost its occult flavor and become every sailor's everyday tool, it came out onto the open deck. Still, in Columbus's day, a pilot who used the magnetic compass might be accused of trafficking with Satan."

The Chinese were the first, perhaps around AD 1000, to use a magnetic needle for navigational purposes. While the strange power of the lodestone was known in both the East and the West, the Chinese were the first to master this phenomenon and to invent the compass. Four hundred years later, the Chinese were sufficiently well-versed with compass and seagoing navigation to mount ambitious sailing expeditions under a famous admiral, Zheng He, who was known as the Three-Jewel Eunuch, perhaps because he gave gems as gifts. With some sixty large ships capable of carrying thirty-seven thousand men, Zheng He carried the message of Chinese superiority to the trading nations of the Indian Ocean. By 1431, Zheng He had established China's preeminence throughout the southern seas, all the way to India, Arabia, and East Africa, well before Europeans seriously attempted voyages to this part of the world. Ironically, Zheng He's seventh voyage was his last. Court politics put an end to foreign adventures and China, the nation that built the Great Wall to keep out foreigners, once again turned inward. Rejecting a spirit of agressive inventiveness and exploration, the xenophobic Chinese cut short an exploring era that might have changed the course of history had it been permitted to flourish.

While it is very possible that the compass idea found its way from China to Europe, it seems just as likely that the European world made the discovery on its own, only a bit later. One legendary account gives credit for the mariner's compass to an anonymous Italian sailor of the early fourteenth century.

The point is that somebody got the notion of attaching a pivoting, magnetized needle over a "compass card," a circle divided into the principal points of direction. At first the sailor's compass card was circular and was kept lying flat on a table. In a dish beside it, a bare magnetic needle was floated on a piece of cork or straw. As the needle pointed north, the card was adjusted to indicate direction.

Before the introduction and widespread use of the magnetic compass, navigators had to rely on an ancient technique called dead reckoning. Basically, the sailor used his experience, instincts, and whatever local knowledge there might be in combination with astronomical sightings. Unfortunately, with dead reckoning the emphasis too often was on "dead." It was a dangerous and unreliable method that presented an enormous obstacle to long seagoing voyages into vast, uncharted expanses of ocean. Certainly, without the compass, Columbus could not even have considered attempting his ambitious dream of reaching the Orient by sailing west.

Why Didn't the Chinese, the Africans, or the Arabs "Discover" America?

As the exploits of Zheng He, the Chinese "Christopher Columbus," and the Chinese invention of the compass demonstrate, Europe was far from alone in making navigational progress and other geographic advances. Seafarers from the Orient and Arabic countries were roaming vast expanses of the eastern oceans while Europe was still in the midst of the Dark Ages.

In fact, Europe's Dark Ages were a period of extraordinary mathematical and scientific advancement in non-European societies. While the medieval church either burned or buried the classics of antiquity and scholars fixed their sights on locating paradise, the Arabic world was adopting Greek notions and embarking on its own golden age of scientific and mathematical advancement. While Ptolemy was neglected by the European world, the Arabs were perpetuating the study of astronomy and mathematics and rapidly expanding their empire in the intellectual vacuum created by the demise of Rome.

But the question remains why these societies, for all their skill and scholarship, and certainly possessed of technical abilities, failed to embark on the type of oceangoing exploration and colonization that would alter the course of history in the late fifteenth century—the European "age of discovery."

Presuming that none of these cultures did in fact reach the Americas first, this question gets to the heart of the relationship between geography and destiny. What geographic or cultural factors kept the Arabs and Chinese (in particular) from reaching out across the oceans to discover the New World?

For the Chinese, it may simply have been the reluctance of a civilization content with what they had to make the enormous effort and sacrifices that discovery and exploration demanded. Daniel Boorstin has labeled this civilization "an empire without wants." Tradition also played a role. This was a people whose cultural and historical tradition had been to oppose or at least limit contact with foreigners. After all, they were the heirs to the people who began to build the Great Wall in 214 BC to keep out foreign marauders.

But there is another, controversial side of the story. In 2002, Gavin Menzies, a retired British submariner, published a book called *1421: The Year China Discovered America*, which became an international bestseller. In it, he claimed that the Chinese admiral Zheng He had sailed to America and produced maps, which were later used by Europeans. Largely dismissed by scholars as an interesting fiction, and lacking archeological evidence of such an astonishing event, the book still enjoyed international success. In 2010, *National Geographic* announced that the so-called Chinese Columbus map was likely a fake.*

*http://news.nationalgeographic.com/news/2006/01/0123_060123_chinese_map_2.html

The Arabs of the medieval period similarly had the means to sail the world. Certainly, they were more advanced than their European counterparts in many respects. Instead of burning the so-called pagan texts of writers like Ptolemy, they studied them and improved upon them. But in more practical terms, the Arabs lacked the fundamental reason that later Europeans like Columbus had for their voyages. They didn't need to find a way to sail to the East. They were already well established there and had little interest in expanding their contacts with Europeans who had shown their colors during the crusades.

That raises a third point of difference. Unlike European explorers who carried the cross as well as the flag of their sponsors, there was no missionary zeal among the Chinese. And while the Arabs took Islam with them, neither culture produced the equivalent of the Jesuits or any of the other Roman Catholic orders bent on proselytizing and converting the "heathen." However misguided those Christian efforts might have been, the urge to spread the faith was a powerful force behind the European voyages of discovery.

Geographic Voices   From Marco Polo's Travels (circa 1299)

In this island of Cipangu (Japan) and the others in its vicinity, their idols are fashioned in a variety of shapes, some of them having the heads of oxen, some of swine, of dogs, goats, and many other animals. Some exhibit the appearance of a single head with two faces; others of three heads, one of them in its proper place, and one upon each shoulder. Some have four arms, others ten, and some an hundred, those which have the greatest number being regarded as the most powerful, and therefore entitled to the most particular worship.

When they are asked by Christians wherefore they give to their deities these diversified forms, they answer that their fathers did so before them. "Those who preceded us," they say "left them such, and such we shall transmit them to our posterity."

The various ceremonies practiced before these idols are so wicked and diabolical that it would be nothing less than an abomination to give an account of them in this book. The reader should, however, be informed that the idolatrous inhabitants of these islands, when they seize the person of an enemy who has not the means of effecting his ransom for money, invite to their house all their relations and friends. Putting their prisoner to death they cook and eat the body, in a convivial manner, asserting that human flesh surpasses every other in the excellence of its flavor.

It is to be understood that the sea in which the island of Cipangu is situated is called the Sea of Chin, and so extensive is this eastern sea that according to the report of experienced pilots and mariners who frequent it, and to whom the truth must be known, it contains no fewer than seven thousand four hundred and forty four islands, mostly inhabited. It is said that of the trees which grow in them, there are none that do not yield a fragrant smell. They produce many spices and drugs, particularly lignum-aloes and pepper, in great abundance, both white and black.

It is impossible to estimate the value of the gold and other articles found in the islands; but their distance from the continent is so great, and the navigation attended with so much trouble and inconvenience, that the vessels engaged in the trade . . . do not reap large profits.

Born in Venice in 1254, Marco Polo came from a well-heeled merchant family. At seventeen, he set off with his father and uncle on a diplomatic mission from Pope Gregory X to the court of Kublai Khan, first Mongol emperor of China. Their journey to China took three and a half years through Persia and Afghanistan. After arriving in Peking (Beijing), the young European attracted the notice of the emperor and remained in his service for twenty years.

After returning to Venice, Marco Polo served on a Venetian ship during one of the regular wars between his city and its rival, Genoa. He was captured and imprisoned. During his time in prison, he dictated an account of his experiences in the Far East to a fellow prisoner, relying on notebooks and perhaps a selective memory as well as some degree of imagination. But he was supposed to have remarked, "I have not told half of what I saw."

Although there had been contact between Europe and China for centuries, Polo's account, translated into many languages, whetted the

Europeans' mercantile appetite for expansion and domination of this trade with the East. It was that trade that gave the impetus to the Portuguese to find a sea route around Africa, as well as to Christopher Columbus's ambition to find a western sea route to the vast storehouse of riches that Marco Polo had described.

Who Did "Discover" America?

Picture a day on the moon in the year 2250. (Please realize that this is not far off from 2013 in the spectrum of history dealt with in these pages.) The Japanese-German consortium's lunar colony is preparing to celebrate the bicentennial of the founding of the colony. There will be big parades and huge sales at the interstellar mall, all in commemoration of the first earthlings to land on the moon, in the year 2050.

But a small band of sign-carrying dissidents—members of the menial American drone class—is protesting the event. They want the colony to recognize the first people to land on the moon, the Americans who did so way back in 1969. They are dismissed as a band of crackpots by the mass media who note that even if Americans had landed on the moon—a preposterous notion given their low social standing in the year 2250—they started no colony and left no other indelible impression of having been to the moon.

Five hundred years after Christopher Columbus sailed from Cape Palos, Spain, and into the pages of myth and history, his name excites extraordinary passions. The recent celebration of the five hundredth anniversary of Columbus's first voyage in 1492 reopened a boisterous debate over Columbus's accomplishment and his rightful place in the history books.

But even to use the word *celebrate* in the same breath as the name Columbus invites controversy. One Native American group received substantial press coverage when they celebrated Columbus Day in 1991 as the anniversary of the last year *before* Columbus arrived in America. In Berkeley, California, October 12 is officially no longer Columbus Day. City officials there have instead determined to call it Indigenous People's Day in honor of the societies that flourished before Columbus arrived. (Somehow it's tough to think about running down to the department store for the Indigenous People's Day coat sales!)

Just about everyone, save for a few holdouts like the Knights of Columbus and the Sons of Italy, agrees that Christopher Columbus did not discover America, as generations of children schooled in a folktale version of history once believed. Much of the legendary Columbus story came from the imagination of the famous American teller of tales, Washington Irving. But there is still considerable confusion about what exactly Columbus did discover and whether someone else "discovered" the New World well before Columbus.

Some historians have argued that yes, they did. It is possible, judging from a ship's bell found off the coast of California, that Chinese or Japanese boats, driven by winds, could have reached the shores of America. And ancient Roman coins have been found in South America, again presumably left by sailors on a lost, wind-driven vessel. Perhaps the most famous of the pre-Columbian discovery theories came from Thor Heyerdahl of *Kon Tiki* and *Ra* expedition fame. A Norwegian anthropologist, Heyerdahl has contended that sailors from Egypt or Phoenicia reached South America or Mexico and built the pyramids there.

Another scholar who has argued for an African discovery of America is Professor Ivan Van Sertima. In his book *They Came Before Columbus,* he primarily relies upon artifacts and sculpture from South and Central American cultures that resemble African counterparts to support his theory. While intriguing, these theories have problems. The first of these is the possibility that different cultures make the same developments on their own. Beyond that, it seems obvious to wonder why only selected parts of these cultures were transferred. If the Phoenicians truly came to America, why didn't they bring their alphabet and the wheel, neither of which was in evidence when the Spanish arrived.

Was Columbus the first? Absolutely not, if we take into account the Vikings, who touched base in or around Newfoundland in 1000, set up housekeeping, found the natives not too friendly, and headed home a few years later. So yes, the Vikings were in America before Columbus. But their presence had no lasting impact on either the history of America or the rest of the Western world. The "discovery" of America by Christopher Columbus as the standard-bearer for Europe signaled the opening of an extraordinary period of discovery, colonization, and yes, exploitation, unequaled in human history. It was a period that would not have taken place if not for other extraordinary events taking place in Europe: the shaking off of the dark shrouds of blind faith that kept Europe in the Dark Ages; the celebration of the human mind that became the Renaissance; the slow, torturous move into a new era of rational thought that crested in the Enlightenment. Of course, all that glory has its dark side, most notably the subjugation of the people the Europeans found when they arrived in America and their introduction, not long after their arrival, of African slaves and the great Atlantic slave trade.

Columbus's once heroic sheen has been considerably tarnished. The history books now tell us that he was sent back to Spain in chains for his grievous mismanagement of the colonies he had started. But even now the myths remain powerful. There are still people who cling to the notion that Columbus was sailing out into a completely unknown world, with a fearful crew believing the voyage was doomed because the earth was flat and they would simply sail off the edge.

In a word, this is simply nonsense. While the conception of the world in the later fifteenth century was still largely shaped by legend and mythology, Columbus certainly knew the world was round, as most rational people did, since it was an idea dating back to the great Greek philosophers.

There are still many who don't realize that Columbus never reached the shores of what is today the United States of America. His voyages took him around the Caribbean and to the coast of South America. He went to his deathbed thinking he had arrived at some island outposts of China, his original goal.

There were also millions of people living in the Americas when Columbus arrived, spread across two continents. It seems difficult to concede that someone has "discovered" a place where forty million people already live! It's like telling your friends you've discovered a great new restaurant. Great hamburgers and french fries. You're just not crazy about the decor, with those golden arches.

The obvious answer to the question of who discovered America is that it was discovered by the people already living here when Columbus dropped anchor off present-day Haiti. They had begun to arrive in America fifteen to thirty thousand years ago, crossing the land bridge that connected Asia and America during one of the last ice ages. In successive waves they came following the game they hunted, eventually settling down and spreading out across two vast continents.

The one constant in the discussion of human views of the shape of the world and its place in the universe is that it has been an ongoing process of change. So while it is easy to snicker at earlier people's notions of the world with their mystical beasts, flat earth floating on an endless sea, and dung-rolling beetles, it is important not to get too secure in our own notions of the world.

The universe has a sneaky way of walking up to us and saying "You think you're so smart, but you've got it all wrong." It happens whenever science uncovers something that shakes our understanding of the world all over again.

For instance, if you went to grade school twenty years ago or more, you can take pretty much everything you learned back then or think you remember about dinosaurs, man's evolution, or outer space, and throw it out the window.

During the past decade alone, most of what we learned and accepted as scientific fact a mere twenty or thirty years ago has now either been radically revised or flatly refuted. Such basics as the age of mankind and the makeup of the universe have gotten major rewrites because of significant discoveries in recent years.

One of the most serious popular misconceptions about science and scientists is that these folks with white jackets and lots of pens in their shirt pockets know it all. Any self-respecting scientist or engineer will be the first to dismiss that notion with utter conviction.

Here is one notable example of how scientific advances can shake up the way we think. Look at Christopher Columbus again. You may have laughed when you read the words that open this chapter about his idea that the world was shaped like a pear.

Well, the joke's on you.

One of the things we learned when the *Vanguard* satellite, launched in March 1958, went into orbit was that the earth is not a sphere, as we learned in grade school—a "fact" we have accepted since the days of Newton. In fact, astronomer John O'Keefe determined from *Vanguard*'s orbit that the earth is slightly pear-shaped, with a bulge in the Southern Hemisphere—though not to the extent Columbus imagined.

But give him credit. At least on this, Columbus was basically right!

Geographic Voices   From *The Tempest*, by William Shakespeare (1564–1616)

> O wonder!
> How many goodly creatures are there here!
> How beauteous mankind is! O brave new world,
> That has such people in it.

Spoken by Miranda, the beautiful daughter of Prospero, when she first sees a young man, these lines are from a play that is a wonderful example of the intersection of geography, history, and art. *The Tempest*, one of Shakespeare's last plays, was written around 1611. It is the tale of Prospero, a duke who has been overthrown and banished to an island where he rules with magical powers. When the men who overthrew him pass near the island, Prospero raises a storm that wrecks their ship and casts them ashore.

The source of the play was an actual incident, widely publicized in England at the time. An expedition of nine ships bound for the newly established colony of Virginia set out from England in May 1609. The flagship, *Sea-Adventure*, was wrecked near Bermuda. All on board were saved, however, and built new ships on the island, eventually reaching Virginia a year later. The accounts of their adventures were sent to England and published.

Shakespeare wasn't the only one to find inspiration in such a notable shipwreck tale. A little more than one hundred years later, Daniel Defoe (1660–1731) published his famous novel *Robinson Crusoe* (1719). Like *The Tempest*, the story of a castaway, Defoe's tale was based on the actual experiences of Alexander Selkirk, a Scottish seaman. While serving on a privateering expedition in 1704, Selkirk was set ashore at his own request on Más a Tierra, one of the Juan Fernández Islands in the Pacific, four hundred miles west of Chile. He was rescued in 1709 and returned to England, where he became a celebrity and achieved literary immortality when Defoe based his timeless story on Selkirk's strange adventures.

In fact, the period of the great European age of discovery inspired a number of other literary classics. One of them introduced a new word into the English language: Thomas More's *Utopia* (1516). Inspired by the tales of wonder being brought to Europe every day with each arriving boat from the new worlds across the Atlantic, More used the form of a traveler's tale to set out his vision of a perfect, altruistic society. In Utopia there was no private ownership. In this highly democratic republic located somewhere off the coast of South America, the most worthy goal for the people was the "leisure" to pursue learning and self-improvement. (The irony of this ideal society free of poverty and suffering is that the word utopia literally means "no place"—More's little joke.)

The other two most notable literary "travel books" influenced by the extraordinary discoveries and explorations of this period are Sir Francis Bacon's *The New Atlantis* (1627), which, like *Utopia*, posits an ideal society on a Pacific island, and Jonathan Swift's *Gulliver's Travels* (1726). Like the others, Swift relied upon enormous public fascination with newly discovered exotic places to craft his satire of English and European manners and society in his tales of a traveler to the islands of Lilliput, Laputa, Brobdingnag, and a variety of other remote nations of the world.

Milestones in Geography I
5000 BC to AD 1507

Before Christ

c. 5000 Sailing ships are used on rivers in Mesopotamia.

c. 3000 Sailing ships are used on the Nile in Egypt.

c. 2900 The Great Pyramid of Giza in Egypt is built as a tomb for Pharaoh Cheops (Khu-fu). The sides of the pyramid, an almost perfect square at the base, are aligned on exact north-south and east-west lines.

c. 2400 The Chinese introduce a method of taking astronomical observations based on the equator and the poles; although not adopted in the West until the sixteenth century AD, this remains today the standard astronomical recording method.

c. 2300 A map of the city of Lagash in Mesopotamia is carved in stone in the lap of a statue of a god; it is the oldest known "city map."

c. 2000 The first use of sail on seagoing vessels in the Aegean Sea.

c. 1800 Under Hammurabi, the famous lawgiver, star catalogs and planetary records are compiled in Babylonia.

c. 1350–1251 Moses leads the Hebrews in the Exodus from Egypt. (This date is in dispute; other suggested dates are as late as 1225 BC.)

c. 1100 The Phoenicians, occupying what is now modern Lebanon, begin their expansion in the Mediterranean region. Primarily seafaring traders, they will colonize the Mediterranean, most notably in Carthage. Although eclipsed by the Greeks, their contributions were significant, particularly in navigation and in the development of an alphabetic script from which the modern Western alphabet is derived.

c. 900 A Babylonian world map is drawn on clay.

c. 750 Greek city-states begin to expand throughout the Mediterranean.

c. 530 Pythagoras, a mathematician and mystic, is active in Samos. The Pythagoreans teach that the earth is a sphere and not in the shape of a disk. The first recorded description of the earth as a sphere, by Plato, quoting Socrates, comes in 380 BC. One of the later members of the Pythagorean school suggests there is a central fire around which the earth, sun, moon, and planets revolve; he also believes the earth rotates.

c. 500 Greek historian Hecataeus develops a map of the world showing Europe and Asia as semicircles surrounded by ocean.

c. 480 Greek philosopher Oenopides is supposedly the first to calculate the angle at which the earth is tipped. His value of 24 degrees is only half a degree from the presently accepted value of about 23.5 degrees.

5th century BC Chinese astronomers begin continuous star observations.

c. 390 Plato conceives the idea that there must be a continent directly opposite Europe, on the other side of the globe, which he calls the Antipodes.

334 Alexander of Macedon invades Asia Minor; conquers Egypt (332) and Persia (330); reaches India (329); dies in Babylon (323).

c. 310–230 Aristarchus postulates that the earth revolves around the sun, according to Archimedes.

c. 300 Chinese writings contain the first reference to a lodestone's magnetic alignment; it is called a "south pointer."

c. 240 Eratosthenes calculates the circumference of the earth with near accuracy.

221 Shi Huang Ti, of the Ch'in dynasty, unites China, and the construction of the Great Wall begins in 214.

c. 190–120 Hipparchus, a Greek astronomer, is the first to use latitude and longitude.

c. 138 During the Han dynasty, Chang Ch'ien is sent as an ambassador to explore Central Asia in search of allies to help fight the Huns. In his years of travel, he reaches Afghanistan and later as far as Syria and possibly Egypt.

c. 112 Opening of the Silk Road across Central Asia; it becomes the route on which Chinese and European goods are exchanged, although no Europeans would see China for centuries to come.

c. 64– 21 Strabo, influential Greek traveler and geographer.

Anno Domini

23–79 Pliny the Elder, a Roman scholar, whose *Natural History* was a widely accepted source. Careless in his research, Pliny passes along fabulous tales and legends for popular consumption. He died observing an eruption of Mount Vesuvius.

117 Death of the Roman emperor Trajan; the Roman Empire is at its greatest extent.

127–145 Claudius Ptolemy, mathematician, astronomer, and cartographer, publishes his major works in Alexandria.

271 The magnetic compass is used in China.

Late 3rd century The first Chinese silk maps with rectangular grid coordinates appear.

c. 350 The Peutingerian table, a Roman route map, is drawn. This road map, showing the known world at the height of the Roman Empire, is a long, narrow strip measuring one inch by twenty-one inches. Tracing a network of seventy thousand miles of Roman roads, it depicted spas, staging routes, and towns both large and small. But all roads were drawn in a straight line; there was no attempt to depict scale, although distances between certain points were written on the map.

410 Visigoths invade Italy, sack Rome, and overrun Spain.

6th century The time of Cosmas Indicopleustes, a traveler and

Christian topographer whose vision of the world relies on biblical scripture instead of scientific accuracy.

632 The death of Mohammed; Arab expansion begins.

760 Arabs adopt Indian numerals and develop algebra and trigonometry.

800 Charlemagne is crowned emperor in Rome, marking the beginning of a new Western (later Holy Roman) Empire.

c. 1000 Vikings colonize Greenland and "discover" America, establishing a colony in Newfoundland. Their stay in North America is brief and leaves no lasting impact on the history of the continent.

1095 Pope Urban II calls for the First Crusade, aimed at recovering Christian holy places in Palestine from the Muslims. In 1099, Jerusalem is captured by a combined European army that slaughters Jews and Muslims there. The crusaders set up a Latin kingdom in Jerusalem, which falls to Saladin in 1187. Quickly losing sight of religious zeal in exchange for territorial conquest, the crusaders continue to assault the Muslim strongholds through 1270. Although a military failure, the crusades profoundly affect Europe by expanding European contact with the East, stimulating trade and contact with the far more developed cultures of the Middle East, China, and India.

1206 Mongols under Genghis Khan begin their conquest of Asia.

1275 Marco Polo arrives in China, enters the service of Kublai Khan. Polo's account of his experiences in China and the East appears in 1299.

1291 The Vivaldi brothers of Genoa attempt to sail around Africa to the East. Although their expedition disappears, it represents a new adventurous spirit in Europe as well as a rejection of the idea that it is impossible to sail around the tip of Africa.

1374 Death of Ibn Battutah, the greatest medieval Arab traveler, known as the "Arab Marco Polo." His book, *Travels*, describes his extensive journeys throughout the Muslim world, China, and Russia—journeys totaling some seventy-five thousand miles (at a time when there were definitely no frequent-flier bonuses).

Geographic Voices   From *Travels in Asia and Africa* (1354), by Ibn Battutah (b. 1304), writing of his trip to Mali and his stay with Muslim Negroes in the Niger River area

Another of their good qualities is their habit of wearing clean, white garments on Fridays. Even if a man has nothing but an old worn shirt, he washes it and cleans it, and wears it to the Friday service. Yet another is their zeal for learning Koran by heart. They put their children in chains if they show any backwardness in memorizing it, and they are not set free until they have it by heart. I visited the *qadi* in his house on the day of the festival. His children were chained up, so I said to him, "Will you not let them loose?" He replied, "I shall not do so until they have learned Koran by heart."

Among their bad qualities are the following. The women servants, slave girls and young girls go about in front of everyone naked, without a stitch of clothing on them. Women go into the sultan's presence naked and without coverings, and his daughters also go about naked. Then there is their custom of putting dust and ashes on their heads, as a mark of respect, and the grotesque ceremonies we have described wherein the poets recite their verse. Another reprehensible practice among many of them is the eating of carrion, dogs and asses. . . .

We halted near this channel at a large village, which had as its governor a Negro, a pilgrim and a man of fine character named Farba Magha. He was one of the Negroes who made

the pilgrimage in the company of the Sultan Mansa Musa.
Farba Magha told me that when Mansa Musa came to this
channel, he had with him a *qadi*, a white man. This *qadi* attempted to make away with four thousand mithqals and the
sultan, on learning of it, was enraged at him and exiled him
to the country of the heathen cannibals. He lived among them
for four years, at the end of which the sultan sent him back to
his own country. The reason why the heathens did not eat him
was that he was white, for they say that the white is indigest-
ible because he is not "ripe," whereas the black man is "ripe" in
their opinion.

1375 The *Catalan Atlas* is completed by Abraham Cresques. A
spectacular map, commissioned as a gift for the king of France,
the atlas contains tantalizing stories of the riches of Mali in
West Africa, where gold "grew like carrots," was brought up "by
ants in the form of nuggets" and was mined by "naked men who
lived in holes."

15th century Ptolemy's *Geographia* is revived in Europe during
the Renaissance.

1405 The Chinese begin voyages in the Indian Ocean under Admiral Zheng He, the Three-Jewel Eunuch, who is later known
as the Chinese Christopher Columbus for his wide-ranging voyages.

1410 French theologian Pierre d'Ailly writes *Imago Mundi*, in
which the size of the earth is greatly reduced. Columbus reads and
underlines d'Ailly's work, this passage in particular: "The length of
the land toward the Orient is much greater than Ptolemy admits . . .
because the length of the habitable Earth on the side of the Orient is more than half the circuit of the globe. For, according to
the philosophers and Pliny, the ocean which stretches between
the extremity of further Spain (that is Morocco) and the eastern
edge of India is no great width. For it is evident that the sea is
navigable in a very few days if the wind be fair, whence it follows

that the sea is not so great that it can cover three quarters of the globe, as certain people figure it."

1415 The Portuguese capture Ceuta on the North African coast (Morocco), marking the beginning of Portugal's African empire. A few years later in 1444, the Portuguese bring the first African slaves back to Europe.

1453 Constantinople, the capital city of Eastern Christianity, is captured by the Turks, who convert St. Sophia's Basilica into a mosque and end the contact between West and East that had flowed through this city that connects Europe and Asia. But a rush of Greek scholars into Italy helps hasten the Renaissance.

1457 A map of the world drawn by Fra Mauro, a Venetian monk, contradicts Ptolemy's idea that Africa is connected to another large southern continent and holds out the possibility of an ocean voyage around Africa.

1487 Bartholomeu Dias rounds the tip of Africa.

1492 Columbus lands in the New World, although he will always believe he has reached the Asian continent. His voyages into the Caribbean mark the opening of the age of European discovery, colonization, and exploitation of the Americas.

1494 The Treaty of Tordesillas divides the New World between Portugal and Spain.

1497 Italian navigator Giovanni Caboto (John Cabot), sailing under the English flag, reaches Newfoundland and claims it for England.

1497–98 Vasco da Gama becomes the first European to sail to India and back.

1498 Columbus reaches South America during his fourth voyage. Although he realizes he has found a vast continent, he still believes it to be connected to Asia.

1499 Amerigo Vespucci, an Italian navigator, reaches America. In later letters that are widely read in Europe, he takes credit for having discovered a *mundus novus,* or new world.

1500 Pedro Cabral discovers Brazil.

1505 The Portuguese establish trading centers in East Africa.

1507 The Waldseemüller map names the New World after Amerigo Vespucci, not Columbus. Even though the mapmaker later changes this on another map, the name sticks.

Geographic Voices   Amerigo Vespucci in a letter written to Lorenzo Medici

In days past, I gave your excellency a full account of my return, and if I remember aright, wrote you a description of all those parts of the New World which I had visited in the vessels of his serene highness the King of Portugal. Carefully considered, they appear truly to form another world, and therefore we have, not without reason, called it the New World. Not one of all the ancients had any knowledge of it, and the things which have been lately ascertained by us transcend all their ideas. They thought there was nothing south of the equinoctial line but an immense sea, and some poor and barren islands. The sea they called the Atlantic, and if sometimes they confessed that there might be land in that region, they contended it must be sterile, and could not be otherwise than uninhabitable.

The present navigation has controverted their opinions, and openly demonstrated to all that they were very far from the truth.

Born in 1454, Amerigo Vespucci has become a historical mystery figure. Although his name was attached to the enormous lands of the

New World, he was later vilified for his supposed deceit in usurping Columbus. From a successful Florentine family, Vespucci went to work for the powerful and wealthy Medici banking family. He was sent to Spain by the Medicis and became an outfitter of ships. In 1499, inspired by the news of Columbus, he joined an expedition of two ships that sailed to South America. In 1501, sailing for Portugal, he made another voyage, after which he determined that these lands were not part of Asia but a new world. Vespucci's travels became far more famous in his day than those of Columbus, leading the mapmaker to use his name to anoint the New World he was adding to his map.

2

What's So Bad About
the Badlands?

So Geographers, in Afric-maps,
With savage pictures fill their gaps,
And o'er unhabitable downs
Place elephants for want of towns.

—Jonathan Swift
On Poetry, a Rhapsody

How Old Is the Earth, and How Was It Formed?

Were the Continents Actually Attached at One Time?

Is the Earth's Crust Well Done?

Where Are the Tallest Mountains?

Earthquakes: Who's at Fault?

Major Historical Earthquake Disasters

What Are Continents?

Which Is the Largest Continent?

One of the basic problems many people have with geography is that they just don't understand it. They don't speak the language.

Peninsulas and capes. Harbors and bays. Arroyos and aquifers. Estuaries and deltas. We were supposed to learn what all of these things meant back in elementary school. But somewhere along the line our tundras got mixed up with our savannas.

Beginning with this chapter, some of geography's more confusing terms will be introduced, along with an overview of the earth's basic physical features and how they came to be. Starting with the most basic feature of them all.

How Old Is the Earth, and How Was It Formed?

Aristotle was pretty smart, but he didn't know it all. One of his mistakes was the notion that the world had always existed. But he shouldn't feel bad; he wasn't alone. Plenty of other people have just assumed that the earth always was.

Yet almost every culture has produced creation myths to explain the beginning of the world. And until very recently in human history, faith outweighed science in reaching those explanations. Perhaps the most notorious attempt to reconcile the biblical version of creation with known facts was the timetable worked out by Archbishop James Ussher in 1650. In Ussher's chronology of the world, creation took place at 9 A.M. on October 26, 4004 BC. Included in the margins of the King James Version of the Bible for centuries to come, Ussher's version of events was accepted as gospel, literally. Even in modern America, the fighting isn't over. During the 1980s, in several American states, groups of fundamentalist Christians went to the courts in an attempt to require state schools to teach "creationist theory"—a pseudoscience grounded in the biblical version of creation—alongside Darwin's evolutionary theory, in biology textbooks. The creationists even had a powerful ally in President Reagan, whose scientific views must be taken into account. After all, this is the man who once said on the campaign trail that trees were a major cause of pollution! (His press secretary, James Brady, amused reporters by shouting "Killer trees!" as the president's plane flew over a forest.)

Of course, religion held sway for most of history. Science entered the picture very slowly but quickly gathered steam. In 1779, the French naturalist Georges-Louis Leclerc, Comte de Buffon, made the remarkably astute guess that the earth started out hot and was slowly cooling. Believing that the earth was made primarily of iron, he heated iron balls and then measured the rate at which they cooled to come up with an earth age of seventy-five thousand years. Don't laugh. The Comte de Buffon was no buffoon. If wide of the mark in his calculations, he was at least making an attempt to bring scientific methods to the question. He was simply short on good information. Long after de Buffon, other scientists tried similar experiments with different materials.

But with the discovery of radiation early in this century, science got one of its periodic major shake-ups. The previously unknown high heat produced by radioactivity made attempts at measuring cooling metals by simple thermometer readings obsolete. Radioactivity produces temperatures that are off the charts. In time, radioactive dating was discovered, giving science a much clearer notion of the earth's age.

The oldest rocks dated by science so far are about 4 billion years old. Well, that's actually 3.96 billion, if you want to be picky about it.

Found among younger rocks in Canada, these ancient stones—which appear to have gathered no moss—are actually grains of zircon. A brown to colorless mineral, zircon can be heated, cut, and polished to form a brilliant blue-white gem. In Greenland, rock formations have been dated at 3.8 billion years.* Obviously, the earth has to be older

---

*Rocks are dated by measuring the radioactivity of elements in the rocks. During radioactive decay, one element changes into another. By measuring the amount of radioactive element in the rock, and the amount of the element to which it has changed, scientists can tell how long decay has been under way and hence the age of the rock. Zircon's age can be calculated by measuring the extent to which a radioactive form of uranium has decayed into lead, its final decay product. The well-known carbon dating method is limited to rocks containing fossils and other once-living material no more than sixty thousand years old. For older objects, the most commonly used material is potassium-40.

than that, and the best estimate is 4.6 billion years old. That number, by the way, looks like this: *4,600,000,000*. (This is the American usage of billion, meaning a thousand million; in British usage, billion means the much larger number of one million million, or 1,000,000,000,000.)

Billion, trillion, zillion—it all seems pretty much the same. In these days of trillion-dollar budgets being casually discussed in Washington, DC, people often forget how enormous these numbers actually are. When you see the words *million* or *ten million* or *ten billion*, they blur into a simplistic mind picture called Big Number. We have difficulty comprehending the difference between such mind-boggling numbers.

Given that 4.6-billion-year date, modern science assumes the earth was formed at about the same time as the sun. The best guess, theoretically speaking, is that when the sun condensed out of a cloud of interstellar gas, a small amount of material was left spinning outside its main body, like clothes in a washing machine. The pull of gravity brought together materials in what are called *planetesimals*—chunks of rock and frozen liquids ranging in size from a few feet to a few miles across. As these whirling bodies collided, some of them merged in a process called *accretion*. Like the specks of dust that collect into dust balls under your bed, these planetesimals began to come together to form the earth and other planets. As they gradually grew larger through accretion, these emerging planets exerted a stronger gravitational pull, gathering in other smaller planetesimals. But some planetesimals went spinning off on their own, colliding with each other and creating a galactic mess of cookie crumbs that are now called meteorites.

Rocks brought back from the moon by the Apollo astronauts in 1970 proved to be three to four billion years old, younger than the earth and consequently crimping the notion that the earth and the moon were formed at the same time. A new theory of the moon's birth that has caught on among the science guys was originated by William Hartmann and Donald R. Davis and explained by Hartmann in his book *The History of the Earth*. Nicknamed the "Big Splash," Hartmann's idea holds that the moon formed out of a collision between the twenty-million-year-old earth and another large planetesimal that struck the earth a glancing blow. The debris from this cosmic fender-bender was ejected into an orbit around the earth and came together to form the moon, which is made of materials similar to those of the earth's.

Were the Continents Actually Attached at One Time?

What *was* that? Did you hear a bump in the night? Perhaps it was California rubbing shoulders with the Pacific Ocean.

One of the most perplexing theories about our so-called solid earth is the notion that all of the land on earth isn't sitting still but sloshing around like toy boats in a wash tub on which small boys are banging hammers!

This notion is technically known as *global plate tectonics*, an almost universally accepted theory. Although the idea that pieces of the earth are in constant motion and that the continents had once been attached to each other goes back hundreds of years, it was formally put forward first by Alfred Wegener in his 1915 book, *The Origin of the Continents and Oceans*. A German meteorologist and naturalist, Wegener was interested in the seeming alignment of the continents. Like a jigsaw puzzle waiting to be put together, the Atlantic coastlines of South America and Africa looked like they could be snapped together into a fairly neat arrangement. When Wegener learned of the discovery of fossils in Brazil that were similar to those found in Africa, it added fuel to his notion that the two places were once connected. Based on these and other bits of physical evidence, Wegener proposed the idea that the continents had once been a single mass, which later split. He even christened his theoretical giant landmass Pangaea ("all land") and surrounded it by an all-encompassing sea, Panthalassa ("all seas").

Like so many other new ideas in science, Wegener's theory of what he called "continental drift" was largely dismissed in his day, principally because he was unable to explain the forces that could propel such enormous landmasses. Wegener thought that tidal pull might have something to do with the process. A visionary and a hero of science, Wegener died in 1930 in Greenland as he attempted to establish a mid-ice observatory. For decades to come, his ideas were simply dismissed as the left-field notions of a crackpot.

But somewhere in science heaven, Wegener is having the last laugh.

A large body of evidence collected since the 1960s has shown that We-gener was indeed on the right path. British geologist S. Keith Runcorn was one of the first champions of the revised notion of continents being connected. Wegener's theory of continental drift has evolved into what is now called plate tectonics theory. Continental drift is out because it is now known that more than just the continents are on the move. The earth's crust is divided into mobile sections called plates. Some of these plates contain continents, or large parts of them; others carry the sea floor. The plates—Canadian geophysicist J. T. Wilson first used the term in 1965—move over the earth's superheated, molten core, pushed and pulled by convection currents in the molten material generated by the heat of the core. The study of the movements of these large plates is called *tectonics* (from the Greek word *tekton*, "to build"). Think of cooking tapioca pudding. You get it to boiling and you can see bubbles move up to the top. If you put something that could float on top of the pudding into the top, it would start bucking and jostling around the cooking pot. The inside of the earth is like a big vat of bubbling tapi-oca, except that the pudding is magma, the liquefied matter within the earth's core that gets blasted out in volcanoes. Floating on top of the magma are the plates—pieces of the earth's crust.

The plates average from thirty to fifty miles in thickness and move at rates as great as a few inches per year. They may be as large as a few thousand miles across—the North American plate stretches for six thousand miles from the Pacific coast to the mid-Atlantic—or as little as a few hundred miles across. People tend to associate rock with the idea of solidity. It is hard to imagine such massive pieces of rock cruising around the planet's surface. The key to their movement is the flexibility of the upper part of the earth's mantle—the layer beneath the crust—which is partly molten. As the plates "float" on this elas-tic, moving mantle, they play a planetary game of hockey—jostling each other for position, rubbing each other the wrong way, pushing one another down, or occasionally banging straight into each other with dramatic and sometimes catastrophic effects. And they are not finished yet. The world we recognize today will be quite different in, say, fifty million years. (Again, this is a blink of the eye in the geological time frame.) For instance, a good-sized portion of East Africa will probably break off. You can see where it will happen if you stand in the Great

Rift Valley, which extends from Syria to Mozambique. And Baja California will have detached itself from the Mexican mainland.

The Major Crustal Plates and Where They Are Going

African plate: heading southwest, away from Europe
Arabian plate: moving north toward Eurasia
Eurasian plate: moving toward the southeast
Australian plate: heading due north
Pacific plate: moving generally to the northwest
North American plate: sliding west toward the Pacific and
    edging south
South American plate: heading west

Back to Wegener's Pangaea, or supercontinent. The best thinking says that all the earth's land masses were concentrated in one great mass some 220 million years ago. As the plates moved apart, Pangaea was split into two huge continental masses. Laurasia, in the Northern Hemisphere, was made up of what became North America, Europe, Greenland, and Asia. In the Southern Hemisphere, the second supercontinent was Gondwanaland, made up of the future Africa, South America, Antarctica, and Australia. Other scientists suggest that the process may have started out with the two smaller masses crashing together around 300 million years ago to form Pangaea, which then later reseparated. But either way, by 100 million years ago, the current continents were taking shape. Europe broke off from North America to join with Asia, and South America was breaking off from Africa. Then, 65 million years ago, India broke off from Africa and moved north on a collision course with Asia. Australia, severed from Antarctica, moved to its present position.

Is the Earth's Crust Well Done?

In a word, no. The pie is definitely still in the oven. And it will never be done.

Constantly realigning and recycling itself, the Earth is in a constant state of change. Most of the time, these changes are imperceptible. But then San Francisco's Candlestick Park starts to rock and roll during the middle of the 1989 World Series and everyone is reminded that Mother Nature has her own agenda. We realize terra firma can be more terror than firma!

More extraordinary images of the changing earth have come recently in pictures from Hawaii and the Philippines, where volcano activity has been dramatic and destructive. Yet while molten lava has flowed down and destroyed plenty of expensive homes, it has also poured into the sea and quickly hardened, adding inches to the coastline and sometimes creating new Hawaiian beaches instantly, which might someday fetch a pretty penny in the real estate market. The trick is managing to stay around long enough to capitalize.

Changes other than the sudden cataclysm of earthquakes and volcanoes are less visibly apparent. We can't see videotape on the evening news of the ongoing process of destruction and creation happening beneath the earth's surface, for instance. Eventually, the changes produced by plate tectonics will be more radical because the continents are still on the move, continuing their leisurely voyage across the mantle of the earth. There are several ways in which the moving plates continue to shape the earth, all of them taking place at spots called plate boundaries, where plates come together. There are three kinds of boundaries. At *neutral*, or *transform*, boundaries, plates rub against each other as they laterally move past each other. The people of California are intimately aware of this action because it causes the earthquakes they feel (and fear) as California is dragged north by the plate beneath the Pacific Ocean.

The second type of boundary is called *diverging*, for the spots where plates are moving away from each other in opposite directions. This is most common under the ocean, as plates carrying the sea floor head in opposite directions. As this happens, magma, the molten material from beneath the earth's crust, wells up and forces itself between the two plates, building massive, submerged ocean mountain ridges. The longest mountain range on earth is the mid-Atlantic ridge that marks the plate boundary between the North American and Eurasian plates, which are heading away from each other. Sometimes this eruption moves above the surface. In 1963 such an immense upheaval occurred near Iceland, where a new island, Surtsey, was born as 10.5 billion feet of molten lava rose out of the sea. But divergent boundaries also exist under land masses; the best example is Africa's Great Rift Valley, where the continent is literally going to be torn apart as the plates go their separate ways. Don't worry. This will take a while and at the present rate, none of us or our descendants will be around to see it.

Finally, there are *convergent* boundaries marking plates meeting head-on. Usually, one of the plates at these boundaries is pushed under the other in a process called *subduction*. The material in the plate getting pushed down is then melted by the intense heat of the mantle, the chief way in which the earth's materials get recycled. If the plates that converge are sea floor plates, one result of the subduction process is a deep oceanic trench, such as the Mariana Trench located in the Pacific, south of Japan, which at more than 35,000 feet (11,022 meters) deep is the deepest spot in the world's oceans. If one of the plates converging does carry land, the result may be a "crumpling" that forms a long mountain chain with a nearby ocean trench. The best example of this action is the Andes Range in South America. And if both plates in the collision carry land, the result is even more spectacular, as the two land-bearing plates are welded together. The Urals, a 1,500-mile-long chain in Russia, marks where Europe and Asia collided and were joined. But probably the most extraordinary collision occurred when the Indian subcontinent rammed into and then joined the rest of Asia.

Where Are the Tallest Mountains?

The collision of India and Asia created the world's tallest mountain ranges, which include Mount Everest, the world's tallest peak. Plate tectonics at work is evident in the fact that Everest is still growing at the rather healthy rate of a centimeter—.394 inch—every year, a veritable growth spurt by geologic standards. Although its precise height is disputed, Everest rises to some 29,108 feet (8,872 meters), or about 5.5 miles high, the tallest in the greatest chain of mountains on earth, the Himalayas (Sanskrit for "abode of snow"). A series of three parallel mountain ridges, the Himalayas run for about 1,500 miles, extending from northwest Pakistan and across Kashmir, northern India, Tibet, Nepal, Sikkim, and Bhutan to the borders of China in the east. The three great rivers of India—the Ganges, the Brahmaputra, and the Indus—all begin in the Himalayas.

Ironically, Everest was named by British mapmakers for a man who may have never seen the mountain. From 1830 to 1843, George Everest was the leader of the British team charged with mapping all of India, a task begun in the eighteenth century and continuing well into the nineteenth, and part of the basis for Rudyard Kipling's *Kim*, the first espionage novel. The India survey was an incredible achievement that required extraordinary bravery and sacrifice in the face of the enormous physical difficulties faced by the surveyors. When the survey reached the Himalayas, Englishmen, along with any other foreigners, could go no farther than the Tibetan border, by order of the Chinese emperor. Instead, the British recruited local men who were trained in surveying skills and equipped with measuring chains disguised as prayer beads. These natives were teachers and educated men called *pandits*. The British pronounced it "pundit," giving the English language a new word for "learned man," a connotation which a number of modern political pundits seem to be doing their best to diminish.

For the most part, the British avoided giving European names to the peaks they surveyed, identifying them either by number or by their local names. For Peak XV, as the surveyors initially numbered Everest, an exception was made when its full height was appreciated and it was christened in George Everest's honor. But Tibetans who lived under its massive heights long ago anointed the great peak with the far more poetic name of Chomolungma ("Sacred Mother of the Waters").

The peak of Everest was not reached until 1953, when Sir Edmund Hillary of New Zealand and Tenzing Norkey, a Sherpa guide from Nepal, reached the summit on May 28. (This introduced a second Himalayan-inspired word into the American political lexicon. *Sherpas* are now the advance people who prepare the way for international summit talks between world leaders.) Everest is claimed by the People's Republic of China, which controls Tibet ("Roof of the Sky") and has officially renamed it the autonomous region of Xizang.

Besides Everest, eight more of the world's ten tallest peaks are in the Himalayas. The second-tallest mountain, K-2 or Mount Godwin-

Austen, stands at 29,064 feet (8,858 meters) in the nearby Karakoram Range in northeastern Pakistan. Together the Himalaya and Karakoram ranges hold nearly all of the world's fifty tallest peaks and a large percentage of the world's hundred tallest peaks.

Compared with these, most mountains in the western hemisphere are fairly small potatoes. The tallest mountain in the west is Cerro Aconcagua (22,831 feet; 6,959 meters), located in the Andes Range, straddling the border between Argentina and Chile. The United States doesn't have a mountain among the world's hundred tallest mountains. In the United States, the tallest peak is Alaska's Mount McKinley (20,320 feet; 6,194 meters), which is also known by its Indian name Denali (an Athabasca Indian word for "the Great One"). The sixteen tallest peaks in the United States are all in Alaska. The tallest mountain in the lower forty-eight states is California's Mount Whitney, which at 14,494 feet is about half the size of Mount Everest. In Europe, the highest point is Mont Blanc (15,771 feet; 4,807 meters) near the Italian-Swiss border in the French Alps. Like the Himalayas, the Alps are the result of a continental head-on collision. In this case, it was the African plate converging with the European plate.

Highest Mountain Peaks

Name	Height (in feet)	Range
Everest	29,028	Himalayas
K-2	28,251	Karakoram
Kanchenjunga	28,169	Himalayas
Lhotse	27,940	Himalayas
Makalu	27,790	Himalayas
Dhaulagiri I	26,810	Himalayas
Manaslu	26,760	Himalayas
Cho Oyu	26,750	Himalayas
Nanga Parbat	26,660	Himalayas
Annapurna I	26,504	Himalayas

SOURCE: *Time Almanac 2012.*

One of the first people to make a case for the formation of the mountains from the bottom of the sea was Leonardo da Vinci (1452–1519). A naturalist who viewed art as an attempt to understand nature, Leonardo hiked through the mountains of Italy and wondered about the creation of the Alps. Observing the presence of fossil seashells, he naturally questioned how they could exist at the top of mountains. Conventional medieval wisdom would have attributed these fossils to the biblical flood that covered the earth, but Leonardo's observations weren't chained to a demanding faith in scripture. He wrote in his notebooks that the surface of the earth was once covered by water, and the mountains of the earth had been uplifted from the ocean floors. Over the years, rains stripped away parts of the mountain, leaving rocky crags. Writing around five hundred years ago, his ideas about mountain building and erosion were uncannily close to the truth as it is understood today.

These are the mountains we can see. There are taller mountains on earth but we don't see all of them and tend to discount them. The volcanic Mauna Kea in Hawaii measures 33,476 feet high, taller than Mount Everest. But only a small portion—13,680 feet—is visible above the surface of the ocean.

Geographic Voices   "Because it is there."

George Leigh Mallory attempted to climb Mount Everest three times. During a tour of America after his second attempt, Mallory was repeatedly asked, "Why do you want to climb Mount Everest?"

His famous reply has become an all-purpose reason for daring to accomplish extraordinary things. Mallory was killed making his third attempt in 1924.

Earthquakes: Who's at Fault?

It takes a fairly substantial earthquake to make headlines. When it happens on live television during the middle of the World Series, as happened in the San Francisco–Oakland area in October 1989, it makes very big headlines.

According to the United States Geological Survey, there are several *million* earthquakes rattling teacups around the world each year. Most of these are so small and occur in such remote areas that they go undetected by the most sensitive of seismic instruments (from the Greek word for earthquake, *seismos*, *seismic* means "caused by or related to earthquakes").

Earthquakes are another result of plate movements. As the plates shift, rocks are either compressed or stretched by their motions. As that happens, the rocks store energy, but the stresses are powerful. Like a spring stretched too far or pressed down long enough, the pressure ultimately must be released. The rocks snap and the tremendous energy stored inside of them is released as an earthquake. Earthquakes occur at all three types of plate borders. At diverging plates, underwater earthquakes are relatively small. But colliding plates can produce huge shocks. In 1976, an enormous earthquake struck China, with a sudden surge of the Australian plate the most likely cause. As the following list of major earthquakes shows, China has been prone to such disasters for much of recorded history and the first attempts to study earthquakes were made by the Chinese. The first recorded earthquake occurred there in 1831 BC and after a quake in 1177 BC, the Chinese began to keep regular records of quakes. An extraordinary Chinese scientist-astronomer named Zhang Heng developed the first seismograph in AD 132. When a tremor hit, a ball fell from the mouth of a bronze dragon into the mouth of one of a series of bronze frogs below to indicate the direction of the earthquake's origin.

Instead of having direct, head-on collisions, some plates are sliding uneasily past each other in a lateral motion. The area between these two ships passing in the night are fault zones. In the United States, the most famous of these is the San Andreas fault on the Pacific coast, where the Pacific plate is heading north and rubbing against the Atlantic plate, taking part of the California coast with it. Contrary to popular belief— and perhaps some wishful thinking on the part of anti-Californians— California is not destined to slide into the Pacific Ocean. Instead, it will be dragged, kicking and screaming, to the north. Computer projections plotting the location of the plates in fifty million years, at present rates of movement, put Los Angeles in the vicinity of Anchorage. Just think of it. La-La, Alaska. In the slightly shorter term, the results may be equally interesting, as Jonathan Weiner put it in his book *Planet Earth*. "In 15 million years, Los Angeles, if it still exists, will be a suburb of San Francisco. The Giants and the Dodgers will again be crosstown rivals."

In the immediate short-term future, the outlook is not amusing. The earthquake that struck San Francisco and Oakland in October 1989 was graded at 7.1 on the Richter scale, a major earthquake capable of inflicting serious damage. Most seismologists believe that California is due for a larger earthquake, perhaps in the 8 or 9 range, which is as high as the Richter scale goes. Devised in the 1940s by seismologists Charles Richter and Beno Gutenberg, the Richter scale measures the amount of energy released by an earthquake and the potential for damage.

The Richter Scale	*Effects*	*Average Number Per Year*
under 2	Imperceptible	600,000
2.0 to 2.9	Generally not felt	300,000
3.0 to 3.9	Felt by people nearby	49,000
4.0 to 4.9	Minor shock; slight damage	6,000
5.0 to 5.9	Moderate shock; energy equivalent to an atomic bomb	1,000
6.0 to 6.9	Large shock; can be destructive in populous areas	120
7.0 to 7.9	Major earthquake; inflicts serious damage; recorded worldwide	14
8.0 to 8.9	Great earthquake producing total destruction to nearby communities; energy released is a million times the first atomic bomb	One every 5 to 10 years
9.0 or more	Largest earthquakes	One or two per century

Major Historical Earthquake Disasters

365—Eastern Mediterranean Affecting an area of about a million square miles—encompassing Italy, Greece, Palestine, and North Africa—this earthquake leveled coastal towns, and the huge wave it spawned destroyed the Egyptian city of Alexandria, drowning 5,000 people there.

1556, January 24—Shaanxi (Shensi) Province, China In the most deadly earthquake in history, an estimated 830,000 people were killed.

1692, June 7—Port Royal, Jamaica A center of British colonial activity and a home port to pirates of the Spanish Main, Port Royal was hit by three shocks that destroyed two thirds of the city, killing thousands. (In 1907, the city was hit again and destroyed by fire following the quake.)

1755, November—Portugal One of the most severe recorded earthquakes leveled Lisbon, a crowded port city with a population of more than 200,000. Buildings swayed and then fell as the city was hit by shocks felt as far away as southern France, North Africa, and even the United States. A sea wave reached as far as the West Indies. The series of quakes that hit Lisbon raised some parts of the coast as much as twenty feet and killed 10,000 to 20,000 people (other estimates are as high as 60,000).

1811, December 16—New Madrid, Missouri Although it is called the greatest earthquake to strike the continental United States, because the land was sparsely populated, this major quake produced few casualties. But its effects were dramatic. The ground rose and fell and the earth opened up in deep cracks. Great waves were created on the Mississippi River, whose course was changed in the quake's aftershocks two weeks later.

1897, June 12—Assam, India This Himalayan region was hit by a shock as large as the New Madrid quake, with dramatic results. The Assam Hills were uplifted by about twenty feet.

1908, December 28—Messina, Italy An estimated 85,000 people were killed and the city destroyed by a powerful earthquake.

1915, January 13—Avezzano, Italy This earthquake left 29,980 dead.

1920, December 16—Gansu (Kansu) Province, China Another huge Chinese earthquake, which killed 200,000.

1923, September 1—Japan Three shocks of 8.3 magnitude rumbled across the Kwanto Plain in Honshu, the principal island of Japan. Huge fissures appeared and landslides changed the landscape. Fires started as cooking stoves overturned and set wooden homes aflame, turning the cities of Yokohama and Tokyo into infernos. Most of Yokohama was destroyed, and Tokyo was almost as severely damaged by the earthquake and fires, leaving 1 million homeless and 140,000 dead—about the same number as were killed in the fire bombings of Tokyo in World War II and the atomic strikes on Hiroshima and Nagasaki.

1935, May 31—India An earthquake in Quetta (in modern Pakistan) killed 50,000.

1939, January 24—Chile Thirty thousand people were killed when an earthquake razed 50,000 square miles of countryside.

1950, August 15—Assam, India Twenty thousand to 30,000 were killed in Assam in one of the most violent quakes in modern times. Registering 8.7 on the Richter scale, its energy was compared to 100,000 Hiroshima-size atomic bombs. The quake was accompanied by ear-splitting noises and sharp explosions produced by the collapsing underground rock structures.

1960, May 22—Chile The city of Valdivia was wrecked, while Concepción was destroyed for a sixth time by earthquakes. This quake was accompanied by giant sea waves, and two dormant volcanoes returned to activity as 5,700 people died and 50,000 homes were destroyed. Because of its location along the Andes Range on the South American coast, Chile has been historically one of the countries hardest hit by devastating quakes.

1964, March 27—Alaska The strongest earthquake ever to strike North America hit 80 miles east of Anchorage. It was followed by seismic waves 50 feet high that traveled 8,445 miles at 450 mph. Remarkably, given the power of this quake, the death toll was only 131. A similar quake in a more densely populated area would have been far more catastrophic. Initially measured at 8.5, it was later upgraded to 9.2.

1970, May 31—Peru Fifty thousand people were killed by an earthquake.

1972, December 22—Managua, Nicaragua This Central American capital was leveled by an earthquake of 6.2 magnitude, leaving some 6,000 dead.

1976, February 4—Guatemala Fifteen major earthquakes struck around the world in 1976, one of them this devastating quake that killed 23,000 people.

1976, July 28—Tangshan, China Another of the deadly 1976 quakes hit this northeastern Chinese city without warning. A year earlier, the Chinese had been able to predict an impending quake and evacuate Liaoning. But no similar predictions were made about this quake. The secretive Chinese government allowed little foreign assistance in response to the disaster, but later studies estimated that more than 600,000 died, half the population of Tangshan.

1976, August 17—Mindanao, Philippines An earthquake and the ensuing tsunami killed 8,000 people.

1977, March 4—Bucharest, Romania Fifteen hundred people were killed when this old European city was struck and razed by a quake.

1978, September 16—Tabas, Iran Twenty-five thousand were killed by an earthquake.

1985, September 19–20—Mexico City, Mexico Measuring 8.1 on the Richter scale, the first of two major earthquakes hit the densely populated Mexican capital. A day later the second hit, measuring 7.6, as the devastated city was in the midst of rescue and repair work. Buildings that had been damaged by the first quake simply collapsed. Despite the city's population of more than 12 million, the death toll was a comparatively low 10,000.

1988, December 7—Armenia A quake measuring 6.9 on the Richter scale killed nearly 25,000 and left another 400,000 homeless. This earthquake had major geopolitical ramifications, as Soviet leader Mikhail Gorbachev, touring the United States when the earthquake struck, rushed back to the stricken region and then welcomed international assistance for the first time in Soviet history.

1989, October 17—San Francisco Bay Area Hitting the Bay Area during the World Series, this major earthquake, measuring 7.1 on the Richter scale, killed relatively few people—67—and was responsible for billions of dollars of damage.

1990, June 21—Northwestern Iran An earthquake measuring 7.7 destroyed villages and towns in the Caspian Sea area, leaving 50,000 people dead and 400,000 homeless.

1991, April 29—Soviet Georgia In the midst of the political tremors shaking the former Soviet Union, an earthquake rolled through the mountainous area in Soviet Georgia just weeks after the former republic declared its independence. Communications were destroyed, buildings razed, and approximately 100 people were killed.

Since 1999, the most deadly earthquakes include:

Date	Location	Number of Deaths
1999	Turkey	17,118
2001	India	20,023
2003	Bam, Iran	26,000
2004	Sumatra, Indonesia	227,898
2005	Kashmir, Pakistan	80,000
2010	Haiti	316,000
2011	Japan	28,050

SOURCE: *Time Almanac*, 2012.

Geographic Voices   A description of the crossing and naming of the Pacific Ocean from *Magellan's Voyage Around the World*, by Antonio Pigafetta

Wednesday, November 28, 1520, we debouched from that strait, engulfing ourselves in the Pacific Sea. We were three months and twenty days without getting any kind of fresh food. We ate biscuit, which was no longer biscuit, but powder of biscuits swarming with worms, for they had eaten the good. It stank strongly of the urine of rats. We drank yellow water that had been putrid for many days. We also ate some ox hides that covered the top of the mainyard to prevent the yard from chafing the shrouds, and which had become exceedingly hard because of the sun, rain and wind. We left them in the sea for four or five days and then placed them for a few moments on top of the embers, and so ate them; and often we ate sawdust from boards. Rats were sold for one-half ducado apiece, and even then we could not get them. But above all the other misfortunes the following was the worst. The gums of both the lower and upper teeth of some of our men swelled, so that they could not eat under any circumstances and therefore died. . . .

We sailed about four thousand leguas during those three months and twenty days through an open stretch in that Pacific Sea. In truth it is very pacific, for during that time we did not suffer any storm.

Sailing for Spain, the Portuguese explorer Ferdinand Magellan (circa 1480–1521) set out in 1519 with a Columbus-like scheme to attempt to reach the Molucca Islands in the East Indies by sailing west. His scheme had an angle. He proposed to find a strait at the extreme tip of South America into the South Sea, newly discovered by Vasco Núñez de Balboa (1475–1519) in 1513. With a polyglot crew of about two hundred fifty men and five creaky ships loaded with guns and trading goods, Magellan sailed from Spain, down the coast of South America. As winter was setting in, he stopped on the Argentine coast and had to put down a rebellion by killing the mutinous captain of one of the ships. Magellan also marooned one of the rebels and a priest who had helped plot the mutiny. After losing one ship, Magellan set off again, finally passing into the straits that bear his name, over three hundred miles of the most difficult and circuitous waters in the world, ending in a narrow passage overlooked by ice-covered mountains. Just as he neared the Pacific, Magellan learned that he had been cheated by his suppliers and was missing a year's worth of provisions. A ship dispatched to look for supplies sailed back to Spain. Persisting with three ships, he finally reached the Pacific, which was named for the fine weather they had encountered.

In March 1521, after the privations described above, he reached Guam, where the crews picked up fresh water and food. A month later, while stopping on the island of Cebú, in the Philippines, Magellan and a small party were lured ashore by the local king. On April 27, 1521, Magellan was killed by poison arrows as he held off the natives while his men retreated to their boats. Two more of the boats were finally abandoned before the journey back to Spain. On September 8, 1522—nearly three years after their departure—eighteen men, the remnants of the two hundred fifty crewmen who sailed, returned to Seville under the command of Sebastián del Cano. The world had been proved round. Its true size was now known. And the scope of Columbus's encounter with the New World was now clear to all Europe.

What Are Continents?

With all the drifting and crashing of continents, it's a wonder they've stood still long enough for people to give them names. The breakup of the ancient supercontinents left us with seven so-called continents. They are defined as the large unbroken masses of land into which the earth's surface is divided. But that provides a lot of leeway and provokes some logical questions. Is Europe really a continent? Why isn't India one? How can islands be a part of a continent if they're not connected to the mainland?

From ancient times to relatively recent ones, people acknowledged only three continents: Europe, Asia, and Africa. The next two, North and South America, weren't recognized until after Columbus's voyages. Australia and Antarctica, which existed only in theory as Terra Australis Incognita since the time of Ptolemy, went undiscovered and unmapped by Europeans for centuries. Australia wasn't named and put on the maps until the nineteenth century. Antarctica was discovered in 1820, when Nathaniel Palmer, an American whaling captain, found islands off the mainland, and a Russian admiral named Fabian Gottlieb von Bellingshausen reached the continental mainland of Antarctica in 1821.

Which Is the Largest Continent?

By almost every reasonable measure, Asia is the most significant place in the world today. Putting aside American and European prejudices about their self-importance, the course of the world may very well be determined by what happens in Asia in the next few decades. The picture is not very promising. Faced with an exploding population, environmental disasters, and frequent natural catastrophes, Asia looks at an uncertain future.

Including the Asian portion of the former Soviet Union, Asia measures more than 17 million square miles, nearly a third of the Earth's land surface. The continent covers more area than the continents of North America, Europe, and Australia combined. Asia is not only the largest continent physically, but it has a combined population of more than 2.5 billion people, about half the world's total population. More than 1 billion of those are Chinese. There are seventy-eight Asian cities that have populations of more than a million people each.

Nearly rivaling China in population is India, currently home to nearly 17 percent of the world's population. Sometime in the next century, India will supplant China as the most populous country in the world and by 2100, its population will be 1,631,800,000, according to World Bank estimates. Today, at least 100,000 people in Bombay pay rent for the right to sleep on a small stretch of sidewalk.

The world's tenth most populous country—also one of its poorest—is Bangladesh, where the median age is only sixteen years. Once part of Pakistan, from which it is separated by a thousand-mile stretch of northern India, Bangladesh won its independence after a brutal civil war in 1971. One of the most densely populated areas in the world, and with a maximum elevation of only 660 feet, Bangladesh is threatened by constant flooding, the most dangerous and frequent natural disaster in the country. Three major rivers flowing out of the Himalayas—the Ganges, the Brahmaputra, and the Meghna—meet in southern Bangladesh to form the largest delta in the world, and their monsoon-swollen waters are an annual hazard. The deforestation of the Himalayas, caused by massive cutting of trees for firewood, worsens the problems of the Bangladesh floods. In 1984, a severe flood left a million people homeless. Flooding in 1988 inundated three quarters of the country. Add to Bangladesh's own natural woes the regular cyclones that form in the Bay of Bengal, such as the one in April 1991 that killed 125,000 people and left as many as nine million homeless. Yet ironically, Bangladesh has become a refuge for people from neighboring Myanmar, formerly known as Burma, where a protracted civil war is forcing thousands of people to flee to Bangladesh, overtaxing what is already a desperate situation.

In the Philippines, a collection of about 7,100 islands approximately 500 miles off the southeast coast of mainland Asia, 30 million people (of 60 million total) live in absolute poverty. About 95 percent of the population lives on the eleven largest islands.

Set against these scenes of privation is the other side of Asia. The nations of the Pacific Rim, led by Japan, South Korea, Taiwan, and Indonesia, boast some of the most dynamic economies in the world. And at an even greater extreme are the oil-producing nations of the Asian Middle East, including the states of the Arabian Peninsula, Iran, and Iraq. Although parts of the Middle East are more closely associated with North Africa or Mediterranean Europe for historical, religious, and cultural reasons, these countries all form part of the Asian landmass.

Why Is the Orient Called "the Orient"?

The Orient has long been labeled by Westerners as "inscrutable" and "mysterious." Also mysterious is where the word *Orient*—generally meaning the countries of East Asia—comes from. And what does that have to do with arriving in a strange place and stopping to get "oriented"? Or getting lost and feeling "disoriented"?

Both of these words—the place name "Orient" and the verb "to orient"—come from the same source: the Latin word *oriri*, which means "to rise," and the related *oriens*, which means "rising." Since the sun rises in the east, *oriens* was used in ancient times to denote the direction of the rising sun, the land and regions east of the Mediterranean. *Orient*, meaning the eastern lands, eventually passed into the English language in the fourteenth century from the Old French.

Gradually, the word took on a new meaning related to turning to face the east. Cathedrals, for instance, were built to face east, to be "oriented" toward Jerusalem. But it wasn't until the nineteenth century that the word *orient* was used in the more general sense of ascertaining direction, or, to borrow another compass term, "to get your bearings." Nowadays, when a man fumbles with a road map while lost and says he only needs to get oriented, he is obviously not looking for China.

His sensible wife, of course, just says, "Honey, let's stop in a gas station and get directions."

Milestones in Geography II
The 16th Century

1512 Portuguese explorers António de Abreu and Francisco Serrao reach the Moluccas, or Spice Islands.

1513 After a twenty-five-day trek through the dense rain forests of Central America, Spaniard Vasco Núñez de Balboa (1475–1519) sights the Pacific Ocean and names it Mar del Sur (Southern Sea). But political rivals later accuse Balboa of treason and he is beheaded in a public square along with four of his followers, with their remains thrown to the vultures.

1518 Smallpox, a common though not necessarily deadly affliction in Europe, is introduced to the island of Hispaniola (what is now Haiti and the Dominican Republic). Far more than Spanish guns, horses, or military tactics, smallpox would be responsible for decimating much of the native population, which had developed no natural immunity to the disease.

1519 Believed to be the returning Aztec god Quetzalcoatl, Hernán Cortés (1485–1547) enters Tenochtitlán (Mexico City) and captures Emperor Montezuma II, beginning the Spanish conquest of the Aztec Empire in Mexico. Although Cortés was initially driven out, he returned in 1521 with a larger force and completed the conquest of Mexico, extending Spanish rule into lower California.

1519–22 Magellan's ship circumnavigates the globe.

1524–28 Still searching for a sea path to the Orient, Giovanni da Verrazzano maps the Atlantic coast of North America as he searches for a northwest passage that will be simpler and more direct than sailing around Africa or around South America as Magellan had done.

c. 1525 French physician Jean Fernel is the first to calculate the length of a degree of latitude at very near its accepted length of 110.567 kilometers at the equator.

1532 Spain's Francisco Pizarro undertakes the conquest of Peru's Inca Empire, already devastated by smallpox and civil war. Pizarro captures the Inca ruler Atahualpa, executes him, and conquers Peru.

1533 First report of triangulation, a surveying method using a network of triangles, by Dutch cartographer Reiner Gemma Frisius, whose book *De Principis Astronomiae et Cosmographiae* also points out that longitude can be found by comparing clock time to the sun's position.

1534 French explorer Jacques Cartier (1491–1557) makes the first of three voyages to America in search of the Northwest Passage linking the Atlantic to the Pacific. On the second trip a year later, he sailed up the Saint Lawrence River, hoping it was the passage to China, and landed at the Huron city of Hochelaga, the present site of Montreal.

Geographic Voices   Jacques Cartier on the Hurons

The tribe has no belief in God that amounts to anything; for they believe in a god they call *Cudouagny*, and maintain that he often holds intercourse with them and tells them what the weather will be like. They also say that when he gets angry with them, he throws dust in their eyes. They believe furthermore that when they die they go to the stars and descend on the horizon like the stars. Next, they go off to a beautiful green field covered with fine trees, flowers and luscious fruits. After they had explained these things to us, we showed them their error and informed them that their *Cudouagny* was a wicked spirit who deceived them, and that there is but one God, Who is in Heaven, Who gives us everything we need and is the Creator of all things and that in Him alone we should believe. Also that one must receive baptism or perish in hell.

While Cartier and the later French explorers were positively enlightened when placed against the Spanish in terms of dealing with Indians, Cartier's views are typical of the sense of cultural and moral superiority that all Europeans carried in their dealings with the Indians. It was that sense of superiority that made it that much simpler for successive waves of Europeans to rout the Indians from their ancestral lands, kill most of them, and destroy the Indian way of life.

1538 First world map by Flemish geographer Gerardus Mercator (born Gerhard Kremer; 1512–1594), a student of Frisius's. This map is the first to use the names North and South America. In 1544 during the Counter-Reformation he was declared a heretic, but influential friends spared him death during the Inquisition. In 1569, he published a world map featuring a new type of projection of the round earth on a flat map. The Mercator Projection, featuring straight, parallel lines of latitude, was designed to assist sailors in sailing along a fixed line and became the most widely accepted solution to the projection problem, but it greatly exaggerates some distances and areas of land near the poles.

1543 Polish astronomer Nicolaus Copernicus (born Mikolaj Koppernigk, 1473–1543) publishes *Of the Revolution of Celestial Bodies*, in which he proposes a sun-centered solar system in which the earth rotates daily on its axis. Although written much earlier, the book was withheld by Copernicus for fear of reprisals from the Roman Catholic Church.

1566 Philip II commissions the first detailed map of Spain.

1570 *Theatrum Orbis Terrarum* (Theater of the World) by Abraham Ortelius (1527–1598), first updated collection of world maps since Ptolemy's *Geography*. In the collection's introduction, Ortelius's friend Gerhardus Mercator for the first time uses the word *atlas*, after the Greek mythic hero who supported the world on his shoulders, to describe a collection of maps.

1577–80 English explorer, slaver, and pirate Sir Francis Drake (circa 1540–1596) circumnavigates the globe, the first Englishman to do so. The round-the-world voyage of the ruthless Drake—nicknamed the Dragon by his enemies—came primarily at the expense of the Spanish and Portuguese whose ports he ravaged and whose ships he plundered. When Drake arrived back in England with £160,000 for the queen and a 4,700 percent return on the investment of his backers, he was knighted by Queen Elizabeth on board his ship, despite Spanish protests. The logs of this journey were kept as a state secret and all details of the voyage were concealed. In the new international competition, such types of information were becoming valuable trade secrets.

1583 Matteo Ricci (1552–1610), Italian Jesuit missionary, arrives in China. To win favor with the emperor, he demonstrates European clocks and draws world maps, wisely placing China at the center.

1584 Sir Walter Raleigh (1554–1618), the English explorer and courtier, organizes an expedition to colonize North America on Roanoke Island, inside North Carolina's Outer Banks. Marshy, stagnant, and disease-ridden, it was a poor choice and the settlers returned to England in the following year with Francis Drake. Raleigh outfitted another colony that was also dumped on Roanoke by a hasty captain more interested in pursuing Spanish treasure ships than in establishing a sound colony. Supply ships were also diverted to attack Spanish ships. When relief ships, delayed by the battle with Spain's Armada, finally arrived in 1590, the colony had disappeared and has become known as the Lost Colony. No remains of the colonists were ever found, but speculation has it that they moved north and mingled with friendly Indians.

1588 The Spanish Armada, a fleet of a hundred thirty heavily armed ships carrying twenty-seven thousand troops, is sent by King Philip II to secure the English Channel as a prelude to a Spanish invasion of England. Under the command of the notorious "sea dogs" Drake and Sir Richard Hawkins, a much smaller English fleet drives the Spanish away, and many of their ships are lost in a storm as they try to escape back to Spain. The battle marked the beginning of Spain's decline and England's rise as the world's predominant sea power.

1589 English geographer Richard Hakluyt (1552–1616) publishes the first edition of his masterwork, *Principal Navigations, Voyages, and Discoveries of the English Nation.* An ordained minister and teacher, Hakluyt was absorbed by the exploits and adventures of the great British sailors, such as Hawkins, Drake, Raleigh, and Frobisher. The book was a patriotic call to the English to colonize the new worlds. Hakluyt was a hawker—a real estate promoter enthusiastically describing the rich and abundant new lands, starting England on its course of empire building.

Why Is Africa Called the "Dark Continent"?

Few places in the world have been more mythologized or misunderstood than Africa. The modern history of Africa, its development, and its place in the modern world can all be traced to its colonization and exploitation by the major European powers, begun by the Portuguese in the 1400s. Africa was only "dark" to Europeans whose knowledge of sub-Saharan Africa was limited to medieval myths of fantastic beasts and strange people and legends of lands rich in gold. Until the Portuguese began their tentative explorations down the west (Atlantic) coast of Africa, the sub-Saharan Africans living along the coast were protected by the vast expanses of the Sahara Desert to the north and the ocean to the west. No navigable rivers allowed penetration into the interior. Africa developed rich cultures, highlighted by such empires as those of Mali, Songhai, and Kush. Contact with Arab and Chinese sailors and traders on Africa's east, or Indian Ocean, coast had existed for centuries and was far more developed—the chief reason that Islam was well established in many parts of Africa long before nineteenth-century Christian missionaries began to set up shop.

To the Arabs, Africa was not so dark. The Mali Empire, centered in the capital of Timbuktu (Tombouctou) was founded in the 1200s and flourished for the next two hundred years as an Islamic religious and trading center. Unfortunately, it was also a center for the slave trade as well, which existed among nations within Africa long before the first Europeans arrived and began the mass deportation of black Africans to the West Indies and the Americas.

A vast landmass straddling the equator and extending 5,000 miles from top to bottom and 4,600 miles from east to west, Africa is the world's second-largest continent. Its area of 11,677,239 square miles (about 29.8 million square km) is equal to about 20 percent of the world's land area. Curiously, it is the only continent with land in all four hemispheres: northern, southern, eastern, and western. The only place it meets another landmass is the tenuous connection to Asia at the Isthmus of Suez in Egypt. Although English is widely spoken, and Swahili is the most prominent native language, there are more than a thousand separate languages spoken in Africa. Fifty of these are considered major languages, used by more than a million people.

In modern Africa, drought, soil erosion, deforestation, and a population explosion are combining to produce one food crisis after another. The African AIDS crisis dwarfs the problem in other countries; some estimates envision a quarter of the African population eventually being infected. These desperate situations are worsened by the instability of governments that spend inordinate amounts of money on military hardware. Civil wars, as well as the last vestiges of superpower jousting, have added to Africa's woes. In 1986, at the end of the well-publicized African drought that led to the "We Are the World" charities, ten of Africa's thirteen worst-affected countries had suffered from some form of war, civil strife, or a massive influx of refugees fleeing such conditions.

An estimated 300 million people live in the forty countries of sub-Saharan Africa and another 300 million live in the North African countries. But by the year 2025, Africa is expected to have a population greater than the combined populations of Europe, North America, and South America, barring a possible AIDS-related population decline. The population of the East African country of Kenya increases at the rate of about 4 percent a year, the highest growth rate in the world. By the year 2020, its population is expected to be about 46 million. (By contrast, the United States has a fairly low population growth rate, less than 1 percent annually.) By 2100, Nigeria is expected to be the third most populous nation, after India and China. Other African nations with rapidly expanding populations are Ethiopia, Zaire, Tanzania, and Egypt.

One of the greatest problems for millions of Africans is water. Drought and the desertification of Africa will continue to make feeding the continent a nightmare for the foreseeable future. One answer is desalination of seawater, but it is a costly answer. Africa's first desalination plant was opened in 1969 in Nouakchott, the capital of Mauritania, in northwestern Africa, where the population has burgeoned from 12,000 in 1964 to more than 350,000 today. Drought has lowered the country's water level so much that only the very deepest wells produce. Most of Mauritania's livestock has starved to death.

In the other countries of North Africa, similar situations are often only marginally better. In Egypt, the birthplace of one of the greatest and longest-lived civilizations in human history, the peace treaty with Israel has allowed a shift away from grossly inflated military spending, but its population explosion looms intractably. Crammed into the thin strip of usable land along the Nile, it is a country facing enormous problems. In Egypt, 95 percent of the country's 84 million people live within a dozen miles of the Nile River or one of its delta distributaries. If the Aswan Dam upstream were breached, almost every Egyptian would be drowned within three days. That threat became real when Israel reportedly threatened to bomb the dam, leading to negotiations between the two countries.

NAMES: *Malawi, Mali, Malaysia, and Maldives*

Let's sort the whole thing out. This is a fairly typical problem. Some countries just sound a lot like some other countries. And it can be very confusing. Here are four good examples. Very different countries in different places, but it's difficult to keep them straight.

Malawi It might have been easier to keep things separate if they kept the old name, Nyasaland. But that was a vestige of British colonial rule, and this republic in East Central Africa, about the size of the state of

Pennsylvania, peacefully won independence in 1964. Two of its neighbors are Mozambique to the east and Zimbabwe (formerly Rhodesia) to the south. Africa's third-largest lake, Malawi (formerly Lake Nyasa), covers 20 percent of the country. Although free of the political strife afflicting other African nations, already impoverished Malawi has suffered from an influx of refugees from the civil war in neighboring Mozambique.

Mali Long before a European set foot on the west coast of Africa, Mali was celebrated in myth as a land of gold. The *Catalan Atlas* drawn in Europe in 1375 spoke of Mali as a place where gold "grew like carrots." In fact, there *was* gold in Mali. A small landlocked country in northwestern Africa on the fringe of the Sahara Desert, Mali was once the home of a substantial empire based in the eight-hundred-year-old trading center, Timbuktu. Founded around 1100, it quickly became a commercial center for the trans-Saharan trade because of its proximity to both the desert and the Niger River. The basis of the trade was gold in exchange for salt carried across the Sahara by camel caravan. By the late thirteenth century, it was the boisterous capital of the Mali Empire and was home of a major mosque and Islamic scholarly center. In Timbuktu, trade and learning flourished together as they would soon after in European cities. But so did the Arab-African slave trade, which became one of Timbuktu's most lucrative businesses. In 1468, Mali and Timbuktu were conquered by the Songhai, an African-Islamic empire that reached its peak in the late fifteenth century.

By the midseventeenth century, Mali's glories were finished, as successive invasions destroyed the onetime empire. The French took control in 1896 and Mali remained a colony until 1960 when it joined Senegal to form the Sudanese Republic, later becoming an independent republic. Like many other northwest African nations, Mali faces acute shortages of food as a result of drought. The most pessimistic estimates envision a time when Mali will become uninhabitable as the Sahara Desert continues to creep south.

Malaysia Shift your view over to Asia. In a somewhat curious grouping, Malaysia is an independent federation of states with one part, West Malaysia, situated on the tip of the Malay Peninsula—just below Thailand—a finger of land that juts out into the China Sea; and a second part, East Malaysia, sitting four hundred miles away on the island of Borneo, the world's third-largest island. Rich in petroleum, rubber, tin, and agricultural products, Malaysia enjoys one of the highest standards of living in Asia.

Those resources have made Malaysia a rich prize in the past. The Portuguese, Dutch, and British all had a hand in exploiting Malaysia's wealth. Held by the Japanese during World War II, Malaysia became a British protectorate in the postwar era. After independence from Great Britain, Malaysia was initially joined to the nearby Singapore Islands. But a civil war establishing Singapore's independence in 1965 split the countries. The population is divided between the predominant Malays and a large minority of Chinese origin. Ethnic fighting between the two groups has led to sporadic violence.

While a large part of the island of Borneo is Malaysian territory, not all of it is. Parts of Borneo joined Singapore when it became independent. And a small speck of the island (2,220 square miles) became Brunei, an Islamic sultanate. Most Americans had never heard of Brunei until its sultan, wealthy from the tiny republic's oil reserves, became a contributor to Colonel Oliver North's Contra-aid charity scheme, exposed during the Iran-Contra scandals near the end of Ronald Reagan's second term.

Maldives A collection of nineteen coral atolls in the Indian Ocean, the Republic of Maldives lies about four hundred miles southwest of Sri Lanka (formerly Ceylon). For the residents of the nearly twelve hundred islets that make up the Maldives, global warming is not simply an interesting theory or debate topic. Since none of the islands rises more than six feet above sea level, even a minor melting of the polar ice caps and a resulting rise in sea levels would be catastrophic. Evidence suggests that sea levels have risen four to six inches in the last century. One forecast calls for a further rise of eight inches by 2030. Four tiny islands in the Maldives are already being evacuated because of flooding.

Geographic Voices   From *The True History of the Conquest of New Spain*, by Bernal Díaz del Castillo (1632)

Gazing on such wonderful sights, we did not know what to say, or whether what appeared before us was real, for on one side, on the land, there were great cities, and in the lake ever so many more, and the lake itself was crowded with canoes, and in the causeway were many bridges at intervals, and in front of us stood the great city of Mexico, and we—we did not even number four hundred soldiers! and we well remembered the words and warnings given us by the people of Huexotzingo and Tlaxcala, and the other warnings that had been given that we should beware of entering Mexico, where they would kill us, as soon as they had us inside. . . .

When we arrived near to Mexico, where there were some other small towers, the great Montezuma got down from his litter, and those great caciques supported him with their arms beneath a marvelously rich canopy of green-colored feathers with much gold and silver embroidery and with pearls and chalchuites suspended from a sort of bordering, which was wonderful to look at. The great Montezuma was richly attired according to his usage, and he was shod with sandals, the soles were of gold and the upper part adorned with precious stones. . . .

The Great Montezuma was about forty years old, of good height and well proportioned, slender and spare of flesh, not very swarthy, but of the natural color and shade of an Indian. He did not wear his hair long but so as just to cover his ears, his scanty black beard was well shaped and thin. His face was somewhat long, but cheerful, and he had good eyes and showed in his appearance and manner both tenderness, and when necessary, gravity. He was neat and clean and bathed once every day in the afternoon. He had many women as mistresses, daughters of Chieftains, and he had two great Cacicas as his legitimate wives. He was free from unnatural offenses. The clothes that he wore one day, he did not put on again until four days later. He had over two hundred chieftains in his guard. . . .

For each meal, over thirty different dishes were prepared by his cooks according to their ways and usage, and they placed small pottery braziers beneath the dishes so they should not get cold. They prepared more than three hundred plates of the food that Montezuma was going to eat, and more than a thousand for the guard. . . .

I have heard it said that they were wont to cook for him the flesh of young boys, but as he had such a variety of dishes, made of so many things, we could not succeed in seeing if they were of human flesh or of other things, for they daily cooked fowls, turkeys, pheasants, native partridges, quail. . . . So we had no insight into it, but I know for certain that after our Captain censured the sacrifice of human beings, and the eating of their flesh, he ordered that such food should not be prepared for him thenceforth.

Born in the year of Columbus's first voyage, Díaz was among the first generation of young Spaniards who went off to the New World, lured by tales of gold and riches. In 1519, he joined Hernán Cortés for his expedition into Mexico. With five ships and six hundred men, Cortés landed in Vera Cruz and proceeded to burn his boats to prevent his men from demanding they turn back. Aztec emperor Montezuma attempted to send out gifts that would prevent the Spanish from coming further. But these gifts had exactly the opposite effect. The Spanish party was reluctantly welcomed by Montezuma, who believed Cortés to be the returning god Quetzalcoatl. In doing so, the emperor doomed himself and his people, whom Cortés conquered rather easily despite bloody opposition.

Did God "Shed His Grace" on All of North America? Or Did He Stop at the U.S. Borders?

With an area of 9,360,000 square miles (about 24 million square km), North America is the world's third-largest continent. Comprised primarily of the United States, Canada, and Mexico, it includes Greenland and the Caribbean islands, with the state of Hawaii thrown in for good measure (even though Hawaii legitimately belongs with the rest of the Pacific islands of Oceania). Occupying about 16 percent of the earth's land, and with a total population of more than 530 million people, North America is both the third-largest and third most populous continent.

Like the other great landmasses, it is an area of huge contrasts, spreading from the polar north of Canada and Alaska down to the tropics of Mexico and the Caribbean. First settled at least twenty-five thousand years ago by the Mongoloid people who moved from Asia across the land bridge that was later covered by the Bering Strait, North America gradually became home to the tens of millions of Native Americans—later misnamed "Indians" by Christopher Columbus, who thought he had arrived in the West Indies—who were spread out across the continent when the first Europeans arrived in the fifteenth and sixteenth centuries. It was that arrival—some call it an "invasion"—that sparked the explosive growth of North America from largely a rich wilderness into the most powerful agricultural and industrial area in the world.

Most Americans remain woefully ignorant of the ten neighboring provinces of Canada, the world's second-largest country in land area. Ironically, the country's name derives from a Huron Indian word, *kanata*, meaning "a small village." Since the War of 1812, when American forces attacked and burned York (Toronto), relations between the two countries have remained peaceful, if occasionally testy. Canadians struggle to maintain a national identity in the face of the cultural and economic power of the United States. But the two countries' shared border remains the longest undefended boundary in the world. Linked for the most part by a common English language and heritage—with a significant French Canadian minority begging to differ—and similar political traditions, the two countries form the world's largest trading partnership. Following the lead of the European Economic Community, nearly all economic restrictions have been eliminated between the two countries, and the Free Trade Agreement, signed in 1988 (derided by some in Canada as the "Sale of Canada Act"), ended remaining restrictions by 1999.

Despite Canada's vast physical size, its population is relatively small, about 34 million, with eight out of ten Canadians living within a hundred miles of the U.S. border.

America's history with its southern neighbor, the United Mexican States, has not been so placid, although Mexico—and perhaps the Caribbean and other Central American nations—will also soon enter the North American Free Trade arrangement. As with Canada, few Americans have more than a vague sense of Mexican history, with most of their impressions shaped by racial and cultural attitudes hardened during two hundred years of stormy, paternalistic, and often militaristic relations. Mexico's image in America has been further distorted by Hollywood versions of such events and people as the Battle of the Alamo and the life of the outlaw-soldier Pancho Villa.

Recent archeological discoveries and a new generation of scholars have opened up a radically new vision of Mexico's pre-Columbian past. While American textbooks usually started off with a little something about Cortés and Montezuma and the grisly Aztec human sacrifices—sometimes killing as many as ten thousand victims at a time—little was said about the long history of life in the Americas before the first Europeans arrived. Mexico had been the center of a succession of Indian civilizations for over twenty-five hundred years. The Olmec, Maya, Toltec, and Aztec—or Mexica, as they called themselves—cultures developed pottery making, mathematics, metallurgy, sophisticated calendars, astronomical observations, architecture, hydraulic engineering, urban planning, medical treatments, and complex social structures just as their European and Asian counterparts did.

The capital of the Aztec Empire was a marvel of engineering. In 1325, as protection against powerful neighbors, the Aztecs chose the island of Tlatelolco in muddy Lake Texcoco as the site of their city. Building three raised causeways and two aqueducts that brought fresh water into the city, they drained swampland and built a dike ten miles long to keep out salty waters from another part of the lake. Within two hundred years, it was an imperial city covering almost five square miles with a network of canals and bridges that made Tenochtitlán—which means "Place of the Prickly Pear Cactus"—an American Venice.

But the defining moment in modern Mexican history came in 1519, with the arrival of a few ships carrying about six hundred men under the command of Hernán Cortés. With guns, horses, and most devastatingly, the smallpox virus, the Spanish quickly dominated Mexico, Central America, and South America, wiping out as much as 90 percent of the native population—estimated between 1.5 and 3 million—stripping the land of its gold and displacing centuries-old societies. Aztec pyramids were reduced to rubble, their stones used to build Roman Catholic cathedrals. Mexico was ruled as the Viceroyalty of New Spain for the next three hundred years, until 1810 when the Mexicans first revolted, winning independence, temporarily, in 1821. During the next twenty years, a considerable portion of Mexico's territory was stripped away by the United States, first by the rebellion of Texas and its annexation into the Union in 1836, and later with the annexation of California and much of the American Southwest after the Mexican War in 1845–48.

Under a reformer president, Benito Juárez, an Indian, Mexico disestablished the Catholic Church as a state religion and refused payment of its foreign debts in 1855. The response was an invasion by British, Spanish, and French forces, with the French ultimately seizing Mexico City and taking control of the country, declaring Archduke Maximilian of Austria to be emperor of Mexico in 1863. A succession of rebellions and dictatorships finally led to the revolution of 1910–17. Since 1917, Mexico has been ruled as a constitutional republic with a president elected for a single six-year term. Although a democracy, Mexico's politics have been dominated by a single party, the Institutional Revolutionary Party, for decades. But opposition parties on both the right and the left have recently begun to make gains.

With an area about one fifth the size of the United States, Mexico has a population of more than 108 million (2010) and a growth rate of about 2.3 percent, far higher than Canada's rate of 0.7 percent or the U.S. rate of 0.8 percent. One of the most indebted nations in the world, Mexico remains poor even though it is an oil-producing nation, and recent advances in agriculture have improved its food production. Low labor costs have made high-tech assembly of everything from cars to personal computers a growing source of jobs.

But its capital, Mexico City, is poor, polluted, and congested. One of the worst places to build a city, it is vulnerable to nature's deadliest threats—earthquakes and still-active volcanoes. Yet for five hundred years it has been one of the world's largest cities, continuing to grow in spite of the regular predictions of its doom, with a population of nearly 23 million in 2011.

IMAGINARY PLACES: Where Was El Dorado?

The English navigator and adventurer Sir Walter Raleigh (1554–1618) was writing in 1595 of a kingdom supposedly located somewhere between the Amazon and Peru. The name El Dorado was first applied to a man, later to a city, and then to an entire country—all legendary. The legend arose from a custom of the Muysca tribe, living on the high flatlands near Bogotá, of anointing each new chief with a resinous gum and then covering him with gold dust. The chief would then plunge into a sacred lake and wash off the gold while his people threw in offerings of emeralds and gold.

> The empire of Guiana is directly east from Peru towards the sea, and lieth under the equinoctial line, and it hath more abundance of gold than any part of Peru, and as many or more great cities than ever Peru had when it flourished most: I have been assured by such of the Spaniards as have seen Manoa, the capital city of the Guiana, which the Spaniards call El Dorado, that for the greatness, for the riches for the excellent seat, it far exceedeth any of the world. . . . How all these rivers cross and encounter, how the country lieth and is bordered mine own discovery, and the way that I entered, with all the rest, your lordship shall receive in a large chart or map, which I have not yet finished, and which I shall most humbly pray your lordship to secrete and not suffer it to pass your hands; for by a draught thereof all may be prevented by other nations; for I know it is this very year sought by the French.

The custom had ended long before the Spanish arrived, but the legend survived and grew to mythic stature. When the first Spanish explorers heard of this tale, they named this chieftain El Dorado, "the Golden One." The Spanish explorers eagerly perpetuated the myth of El Dorado because it provided such a simple rationale for continued exploration. They even identified El Dorado as the city of Manoa, supposedly in southeastern Guiana. Much of the Spanish exploration and conquest of South America was a direct result of the legend of El Dorado and the searches it inspired.

Even while El Dorado's allure attracted the Spanish and then Englishmen like Raleigh, a second legend sprung up, perhaps the practical joke of vengeful natives who wanted the white men to go away and enjoyed the sport of seeing their white conquerors go wandering all over the country in search of a fantasy. An even greater fortune awaited the man who could find Cíbola, with its Seven Cities of Gold, which were supposed to exist in the southwest area of North America. The legend attracted Spanish explorers, most notably Francisco Coronado (1510–1554), who took three hundred Spanish cavalrymen and a thousand Indians on a long trek in 1540 through much of the Southwest. What they eventually discovered as Cíbola was a collection of Zuñi pueblos, although the expedition also found the Grand Canyon.

Is America One Continent or Two?

If a continent is a large unbroken land mass completely surrounded by water, why call North and South America different continents? They are clearly connected to each other. Central America—which comprises the seven independent republics of Belize, Costa Rica, El Salvador, Guatemala, Honduras, Nicaragua, and Panama—creates a land bridge between North and South America. And even though this thin strip was often flooded in the past, logically speaking, the two continents are one. But political and historical considerations—especially the fact that the history of Canada and the United States was dominated by the British, while Spain retained its control of Mexico and almost everything south of it save Brazil—often override geographical facts. And no one ever said that geography is a perfectly logical science anyway!

Americans may be poor on the facts about Canada and Mexico, their nearest neighbors, but they are utterly desperate when it comes to South America. For instance, most Americans would be surprised to learn that virtually the whole continent of South America lies east of Savannah, Georgia. With an area of 6,883,000 square miles (about 18 million square km), South America is the world's fourth-largest continent, with almost 12 percent of the earth's land surface. South America's 302 million people live in twelve independent republics and one colonial-era holdover (French Guiana). And even though it seems that the population density is comparatively low, South America is intensely urban, because much of the continent is either inaccessible or can't be farmed because of its two most prominent geographical features, the Andes Mountains and the Amazon rain forest.

The Andes Mountains run for approximately 4,500 miles along almost the entire western, or Pacific, coast of South America, more than three times the length of the American Rockies. Passing through seven of South America's twelve republics—Argentina, Chile, Bolivia, Peru, Ecuador, Colombia, and Venezuela—the Andes are second only to the Himalayas in terms of average height. (The other South American republics are Brazil, Guyana, Paraguay, Suriname, and Uruguay; French Guiana is the last remaining European possession on the continent.) Cerro Aconcagua, the western hemisphere's tallest peak at 22,834 feet (6,960 meters), is in the Andes near Argentina's northwest border with Chile. In Chile, a slender thread of country 1,800 miles long, the Andes Mountains cover one third of the land, making much of the country unfarmable. Chile is home to a large, mineral-rich desert, the Atacama, and also claims the world's southernmost city, Punta Arenas.

Hidden in the Peruvian Andes for almost five hundred years was the mystery of Machu Picchu, once a great Incan city. Lords of an extensive and highly centralized empire, the Incas (also the title of the empire's ruler) held territory that extended 3,000 miles from north to south along a 250-mile-wide corridor from the Pacific coastal plain to the high Andes. Although they lacked the wheel or writing, the Incas were master builders, with an elaborate system of roads and cable suspension bridges that allowed messengers to travel as much as 150 miles per day. The Inca system of terraced farms not only produced ample food but controlled erosion of the soil on the steep mountainside farmlands—techniques that are being reintroduced after centuries of colonial neglect and governmental mismanagement. Inca architects raised fine buildings in the capital city of Cuzco, which meant "navel" in the Quechua language of the Incas, another example of the omphalos syndrome mentioned earlier. Goldsmiths fashioned beautiful objects that immediately caught the attention of the Spaniards who arrived in 1532. Weakened by internal wars, the Incas fell easy prey to the conquistadors, who brought devastating smallpox.

But their greatest building feat may have been Machu Picchu. Perched high up in the Andes, on a mountainous crag that drops steeply on every side, Machu Picchu went undiscovered until the American Hiram Bingham reached it in 1911. In the city, steep stairways lead to granite shrines, marvelously carved stone temples and houses, terraced walls built without mortar, and huge ceremonial stones. Streets, stairways, and plazas were all laid out in perfect harmony with the contours of the mountaintop. With windows placed in temples to permit observation of the midwinter solstice, Machu Picchu was thought to be a sacred city where the Inca lords and their "Virgins of the Sun" went to worship. Even though the Spanish had destroyed almost every other vestige of Incan society, Machu Picchu was found almost intact. But its origins and the reasons for its apparent abandonment remain a mystery.

While some researchers suggest that the site was essentially an estate of an Incan emperor (a *very* elaborate getaway spot), *National Geographic* explorer-in-residence Johan Reinhard believes Hiram Bingham was on the right track more than one hundred years ago. Even if those "sacred virgins of the sun," supposedly beautiful young women who had "been educated to the service of the temple and ministering to the wants of the Inca," were merely an appealing but fanciful legend, Reinhard believes that Machu Picchu was a significant religious site, mostly because of its location. In *Machu Picchu: Exploring an Ancient Sacred Center*, he called it "sacred geography" because the site is built in a mountainous region that held great significance in the Inca culture. In his humorous but insightful *Turn Right at Machu Picchu*, Mark Adams says of this archirectural wonder, "The structures and carvings at Machu Picchu were designed to complement all this sacred geography."

In 2007, Yale University agreed to return to Peru many of the objects and artifacts, including human remains, that Professor Bingham had brought to Yale from his excavations. Now a UNESCO World Heritage site, Machu Picchu was also placed on a list of UN Endangered Sites because of environmental degradation, largely due to the large numbers of tourists it attracts every year.

Located on the northwest coast, Peru is today the third-largest South American republic. Once the principal source of Spain's gold and silver in South America, Peru was largely stripped of its wealth and today its economy struggles. Because of the Andes, only 3 percent of the land is arable, and communication and transportation are also made difficult by the terrain, a perfect example of the negative interaction between geography and a nation's economy. Although commercial fishing is a significant part of the Peruvian economy, overfishing of its coastal waters has caused a steep decline in the catch.

South America's other most extraordinary feature is Brazil's Amazon rain forest and its river. Nearly half of South America is covered by Brazil, and its heavily wooded Amazon basin covers half the country. The largest country in South America and the fifth-largest in the world, Brazil is larger than the contiguous forty-eight American states. But the vast Amazon river basin holds only a tiny population. Ten percent of Brazil's 140 million people live in two cities, São Paulo and Rio de Janeiro, and almost half the population lives in the south-central region, which is responsible for 80 percent of the nation's industrial output and 75 percent of its farm products. One of the world's leading debtor nations, Brazil is in the midst of a major economic makeover that is attempting to break decades of unimaginable inflation rates.

Among its other geographical wonders, South America is also home to the world's highest waterfall, Angel Falls (Salto Angel), in southeast Venezuela ("Little Venice," a name given by Amerigo Vespucci, who was struck by the sight of native huts perched above the coastal waters). Hidden in Venezuela's remote forests, the falls drop 3,212 feet from the side of a twenty-mile-long flat-topped mountain, or mesa, known as Auyán-tepuí (Devil Mountain). That is thirteen times higher than Niagara Falls and more than double the size of the Sears Tower in Chicago, the world's tallest building (1,454 feet; 443 meters). Almost totally inaccessible, Angel Falls can be seen fully only from the air, which is how it was first seen and then got its name. It would be logical to think that the waterfall, unlike the nearby mountain named for the devil, is named for celestial messengers. The image certainly applies, as the white water sails through the air from such great heights. But the falls are actually named for an American flier and prospector, Jimmy Angel, who discovered them in 1935 (and crashed his plane nearby in 1937). The falls were not reached by foot and accurately surveyed until 1949, when an American team confirmed their height.

What's the Difference Between a Rain Forest and a Jungle?

"Out of the Jungle" was the cover-story headline of the *New York Times Magazine* early in 1992. The article reported on the political truce and end of a long civil war in the Central American nation of El Salvador. The former rebels had traded their combat fatigues for three-piece suits. So why didn't the headline read "Out of the Rain Forest"? And what's the difference between these geographic terms anyway?

In a word, marketing. At least that is what language maven William Safire believed. In these days of environmental awareness and ecological political correctness, *rain forest* has displaced *jungle* because it sounds a lot more attractive. As Safire wrote in his *New York Times* column in 1991, "Because a jungle was fearsome, nobody would want to preserve it. But a *forest* has a nice ring to it—there was Robin Hood with his merry men robbing the rich in Sherwood Forest—and the word lent itself to persuasion for preservation. If a pollster asks, 'Is it O.K. to mow down the *jungle*?' the answer will be 'Sure, who needs it?'; if the same pollster asks, 'Do you approve of destruction of the *rain forest*?' the answer will be 'No, it will lead to global warming or a new ice age.' "

But sorry, Mr. Safire. There are some differences. A rain forest typically has a high canopy and very little undergrowth, while a jungle is densely undergrown. That makes quite a difference if you are trying to get through one.

The word *jungle* comes from the Hindu and Sanskrit words for wasteland or desert, which seems odd given our very moist image of a jungle. Only gradually did the word become associated with the now-familiar Tarzan-ish scene of dangling vines, deadly snakes, and incessant monkey chatter.

But in the past few years of heightened environmental passion, *jungle* has been macheted out of the language. And *rain forest* has gone big time. Ben & Jerry's, the ice-cream company from the non–rain-forested state of Vermont, sells "Rain Forest Crunch," using nuts grown in the Amazon rain forest, with a portion of their profits going to rain-forest preservation efforts. Supermarket shelves now carry lines of "environmentally correct" rain-forest candy. Hollywood knows a good thing when it sees it. The 1992 film *Medicine Man* enjoyed a modest success, largely due to interest in rain-forest preservation. The cartoon world has it both ways. The Disney cartoon feature *The Jungle Book* is adapted from Rudyard Kipling's classic of the same name. And in 1992 a rain-forest feature called *Fern-Gully: The Last Rain Forest*, was released with the hope that it would do for rain forests what *Bambi* did for white-tailed deer.

There is another kind of jungle called a "monsoon forest" typical of India, Burma, and Southeast Asia. (For more on monsoons and climate in general, see Chapter 5.) Unlike the tropical (or equatorial) rain forests, where it rains all the time, monsoon forests have two distinct seasons, one wet, one dry. Since it rains only half the time, maybe we should call it the "fifty-percent-chance-of-rain forest."

Whether you call them rain forests or jungles, they are often large but also disappearing. North America's largest tropical rain forest, the Lacandona, is in the southeast Mexican state of Chiapas. Named for a nation of Native Americans presumed to have descended from the Maya, the Lacandona rain forest covered five thousand square miles (an area about the size of Connecticut) only fifty years ago. But since 1970, more than 60 percent of the lush but fragile forest has been lost to development.

The Amazon in South America is the world's largest rain forest. It is an area almost as big as most of the United States and contains more species of plants and animals than any other place on earth. But large tracts—as much as 4 percent each year—of the Amazon tropical rain forest have been burned and cleared for cattle ranches, farms, lumber, and for tax incentives.

"So what?" you ask. Converting the wild jungle into usable land sounds like a pretty good idea. But Mother Nature has different views. It is estimated that the Amazonian forest contains a *third of the planet's trees* and supplies *20 percent of its oxygen*. Getting rid of such a huge part of the planet's "lungs" has an impact way beyond Brazil's borders.

The burning of massive tracts of Brazil's rain forest has another dangerous consequence. The smoke from these huge fires contributes to the buildup of carbon dioxide in the atmosphere, potentially contributing to the "greenhouse effect" that will lead to global warming, a catastrophic rise in sea levels, and the possibility of creating deserts where acres of wheat grow today.

Finally there is the issue of biodiversity, a fancy way of saying there are a lot of living things in the rain forest we don't even know about. The destruction of rain-forest land is killing off an incredible number of species whose value, particularly in the area of medical research, is unknown. Only recently has the issue of species preservation come to the forefront of concern over rain-forest destruction, but it is potentially the most significant reason to preserve the regions that remain.

IMAGINARY PLACES: *Are There Amazons on the Amazon River?*

How did a race of legendary warrior women who the Greeks thought lived near the Caspian Sea end up in South America?

To the Greeks, the Amazons were a race of brave female warriors who cut off one of their breasts in order to be able to carry their shields or to draw their bows with greater ease; their name derives from the Greek word for "breastless." One of the labors of the hero Hercules was to steal the girdle of an Amazon queen. Strabo wrote about them in circa AD 23. And the medieval travel writer John Mandeville also described the fierce Amazons in his *Travels* (circa 1356).

According to these and other accounts, accepted as literally true right through the Middle Ages, Amazonia was an empire of women who did not tolerate the presence of men. Their only contact with the opposite sex was at an annual festival designed to ensure the reproduction of their race. Once in Amazonia, the males were dispassionately used and then kept as slaves or expediently disposed of. Only the female children were kept by the Amazons; the boys were sent away.

The idea that there were Amazons in South America dates back to the Spanish conquest. The first Europeans to see the Amazon River were Spanish conquistadors in 1500 who called it the River Sea. Francisco de Orellana, one of those searching for the gold of El Dorado, led the first European descent of the river in 1541, and it was briefly called Orellana. In Orellana's party was Father Gaspar de Carvajal, a Dominican friar who chronicled this expedition and told of seeing women warriors leading the attacks on the Spaniards' boats.

Father Gaspar wrote:

> We ourselves saw these women, who were there fighting in front of all of the Indian men as women captains. These [women] fought so courageously that the Indian men did not dare turn their backs, and anyone who did turn his back they killed with clubs right there before us, and this is the reason why the Indians kept up their defense for so long. These women are very white and tall, and have hair very long and braided and wound

about the head, and they are very robust and go about naked, but with their privy parts covered, with their bow and arrows in their hands, doing as much fighting as ten Indian men. And indeed there was one woman among these who shot an arrow a span deep into one of the brigantines, and others less deep, so that our brigantines looked like porcupines.

According to the Union of Concerned Scientists, Brazil has made dramatic strides in reducing global warming emissions slowing deforestation and regrowing its tropical forests. Between 2005 and 2010, Brazil nearly met its goal of slowing deforestation. Data from 2009–2010 showed that Brazil's area of deforestation had dropped 67 percent. The cooperation of two countries (Norway and Brazil) has achieved a "reduction in global warming pollution comparable to the reductions that both the United States and the European Union have only pledged to achieve by 2020."*

If these were not the fierce warrior women written and told of for centuries in Europe, they were certainly close to the genuine Amazonian article. Adding weight to the notion that these were the Amazons was the fact that the Indians called the river Amazunu ("big wave"). Later Spanish expeditions failed to turn up any of these extraordinary fighting women, but the name River of the Amazons stuck.

The Amazon is the second-longest river in the world, after the Nile. But far more impressive is its volume. The Nile, for its length, is a leaky faucet compared to the Amazon, carrying less than 2 percent of the Amazon's volume. The Amazon contains more water than the Nile, the Yangtze, and the Mississippi Rivers combined—nearly one fifth of the Earth's running fresh water.

The outpouring of the Amazon River, whose source is in the Peruvian Andes, is so great that the open sea is freshwater for more than two hundred miles beyond the mouth of the Amazon. That is sufficient fresh water—nearly seven million cubic feet every second—to supply two hundred times the municipal needs of the entire United States. At its broadest as it nears the ocean, the Amazon is forty miles wide. Seasonal tides—called the *pororoca*—send its waters running upriver in

---

* www.ucsusa.org/global_warming/solutions/forest_solutions/brazils-reduction-deforestation.html

large waves at thirty-five kilometers per hour. Oceangoing boats can travel the Amazon all the way to the city of Iquitos in Peru.

Who Owns Antarctica?

Take your pick. Nobody. Everybody. Or the seven nations that have staked claims to portions of the great ice sheet covering the world's fifth-largest, and most remote, continent. It is also the coldest, windiest, and driest continent—the last a seeming contradiction because 2 percent of the earth's fresh water is in Antarctica. It just happens to be locked up in the continent's ice cap, which is between six thousand and fourteen thousand feet thick and contains 90 percent of the world's ice.

With no native or permanent population, it is certainly the loneliest continent. And it is the only continent to be truly discovered, since nobody lived there when it was found. Centered on the South Pole and lying almost entirely within the Antarctic Circle, Antarctica (which means simply "opposite the Arctic") has an area of more than 5.5 million square miles (15.5 million square km), equal to 10 percent of the earth's land surface. It has steep mountain ranges and two active volcanoes. But there are no flowering plants, grasses, or large mammals there. Species of algae, moss, lichens, and sea plankton provide food for the rich variety of fish, birds, whales, and seals that live on Antarctica or in its nearby waters.

That was not always true. Researchers have found fossil remains of plant and tree life. More recently, dinosaur fossils have been uncovered, suggesting that Antarctica was not always so inhospitable to life. Dinosaur fossils discovered in 1991 in 200-million-year-old rock near the South Pole prove that Antarctica was once a warmer place. These discoveries also support plate tectonics because these Antarctic fossils are the same as certain African fossils, supporting the theory that the two were once connected as part of Gondwanaland, the great southern supercontinent, not at the bottom of the world but somewhat closer to what is now the southwestern Pacific Ocean. Other geologists have suggested recently that Antarctica and America were once connected even earlier. Pushing the geological time clock back to 500 million years ago, these researchers suggest that South America was wedged like a slice of pizza in between Antarctica and North America.

Although it is possible, and even likely, that Polynesian navigators may have reached Antarctica, the first recorded approach to the continent was that of Captain James Cook (1728–1779), perhaps the greatest sailor and oceangoing explorer in history. In late 1772, on his second great voyage of exploration and mapping, he neared Antarctica, even crossing the Antarctic Circle, without actually sighting the continent. Confronted by impassable ice and frigid conditions that no British seaman had ever faced before—even though it was now the Antarctic summer in January 1773—Cook turned back, having sailed farther south than any other man. A year later, on the same voyage, the intrepid Cook came within one hundred miles of the continent but was again forced back by ice and frigid conditions. Surmising that he was looking at a vast frozen ocean, Cook didn't find Antarctica. But in the course of this voyage, he had dismissed, once and for all, the centuries-old myth of Terra Australis, a great, rich southern continent with which the British hoped to establish trading relations.

The mainland and islands off it were finally reached in 1820 by both an American whaling captain, Nathaniel Palmer, and a Russian expedition led by Fabian von Bellingshausen. Within a few years, its waters were being explored by American and English navigators, including the American Charles Wilkes, who determined that Antarctica was a continent, and Englishman James Ross, who discovered the Antarctic sea named for him while charting much of its coast between 1840 and 1842.

By the beginning of this century, the exploration of Antarctica inspired a dramatic but tragic race to reach the South Pole. Lying about a thousand miles inland, the South Pole is a high, flat table—the coldest, most desolate place on earth. A vista of raging blizzards and lifelessness, the pole was reached by Norwegian Roald Amundsen (1872–1928) on December 14, 1911. His party arrived one month sooner than that of Englishman Robert Falcon Scott (1868–1912), whose ill-fated four-man group got to the pole on January 18, 1912. On their disappointed return trip, Scott's party faced two months of starvation, scurvy, and frostbite, before all died only eleven miles from the next supply station.

Since its discovery, Antarctica has been the focus of competing claims for what may be unreachable riches. Only once has there been actual fighting over the territory. In 1952, British scientists were fired on by Argentine soldiers ordered to prevent the British from rebuilding a destroyed scientific base. They were disputing the Antarctic Peninsula, a long finger of land that reaches up toward the tip of South America, only 800 miles away. The British based their claim on their possession of the Falklands, a group of islands about 450 miles (650 km) northeast of Cape Horn at the tip of South America. (Great Britain's claim to the Falkland Islands, or Islas Malvinas to Argentina, led to a war between the two nations on the islands in 1982. After a brief but fierce fight that left 1,000 dead on the two sides, the British retained possession.) During the 1950s, Australia, New Zealand, France, Norway, and Chile also staked claims to Antarctic territory. The United States, the Soviet Union, Japan, South Africa, and Belgium established research stations by 1959 and were later joined by China, India, and Brazil. Although there is a presumption of substantial mineral wealth and petroleum reserves under the ice and in the seas around Antarctica, the continent's demanding climate, tremendous ice depths, and fragile environment pose huge obstacles to recovering any of that estimated wealth.

Since 1961, Antarctica has been governed by the terms of the Antarctic Treaty, which declared that the continent be exclusively used for peaceful purposes, prohibited military operations, and established the continent as the world's first nuclear-free zone. But that treaty lapsed in 1989 and its provisions are open to negotiation. Twenty countries have already agreed to forbid oil drilling in Antarctica. Australia and France have called for a perpetual ban on mineral extraction. But if somebody finds oil tomorrow and a way to get it out from under the ice, how long will the good intentions last?

In the meantime, Antarctica and the South Pole exist as a rather extraordinary science lab whose miles of icy depths and unique conditions provide an array of interesting clues about the earth's past and the atmosphere above. It has already yielded the distressing discovery of the hole in the ozone layer of the atmosphere. (See Chapter 5.)

Isn't Europe Just a Part of Asia?

This may come as a severe shock to Conservatives in Great Britain's Parliament and to other European traditionalists who have strong views about Europe's cultural superiority. But sorry, folks. You are just another part of Asia. It's true, geographically speaking. Europe, including the British Isles, is simply a large western peninsula of Asia. Many geographers, when referring to Europe and Asia, speak of Eurasia. But political considerations and historical precedents often force geographic realities to take a backseat.

The second-smallest continent, Europe, including the European portion of the former USSR, is estimated to occupy 3.8 to 4 million square miles (10.5 million square km), or about 8 percent of the earth's land surface. An exact figure is difficult to obtain because of numerous offshore islands and disputes over exactly where Europe ends and Asia begins. Both Russia and Turkey, for instance, have one foot on the dock and one foot in the boat, so to speak. But the European states—including less familiar little places like Monaco, Andorra, San Marino, and Vatican City—hold more than 25 percent of the world's population.

An exact count of Europe's separate states has been made somewhat tricky in the past year or two. And it's gotten difficult to tell the players apart without a scorecard. With the unification of Germany and the breakup of the Soviet Union, there are now at least forty separate states in Europe, depending on what you call "states" and what you call "Europe"—and they're not done making new countries yet!

Even as Europe is moving toward a unified system of currencies, an end to protective tariffs, and the opening of borders—making national passports obsolete within the continent—the entire continent has been gripped by new waves of fervent and often violent nationalism. Initially, that nationalism was working itself out with a minimum of bloodshed as the astonishing events in Eastern Europe and later the Soviet Union were played out in dizzying speed. But in several nations of Eastern Europe, Yugoslavia in particular, centuries-old ethnic, religious, and tribal antagonism—suppressed during the years of authoritarian Communist rule—have brought the worst bloodshed to Europe since the end of World War II. The following list breaks Europe up into bite-sized pieces to make it easier to recognize what was once a reassuringly familiar place on the globe.

The "New" Europe

An All-Purpose, Loosely Defined Political
and Geographical Roundup

Scandinavia
Denmark
Norway
Sweden
Finland
Though not actually on one of the Scandinavian peninsulas, Finland is culturally and politically considered to be part of this group.

Low Countries
Belgium
Luxembourg
The Netherlands
In 1958, these three countries formed an economic union called Benelux, a convenient reminder of their names.

The Iberian Peninsula
Portugal
Spain

Alpine States
Austria
France
Italy

Liechtenstein
Switzerland

The Balkans
Albania
Bulgaria
Greece
Romania
The former state of Yugoslavia: the scene of the worst violence of the post-Soviet era, this region—formed after World War II as a socialist republic—fell apart after years of ethnic tensions. In 1992, independence was declared by:
Croatia
Slovenia
Macedonia
Bosnia and Herzegovina

In 2000, Yugoslavia became:
Serbia and Montenegro in 2006 (Montenegro declared its independence)

In 2008:
Kosovo also declared its independence from Serbia, and is recognized by some ninety nations.

Central Europe
Czechoslovakia
Democracy came to Czechoslovakia, which elected the writer and former political prisoner Vaclav Havel as president in December 1990. In 1993, the country split into two states:

Czech Republic
Slovakia

Germany
The new Germany is made up of the former Federal Republic of Germany (West Germany) and the German Democratic Republic (East Germany). In 1999 the capital was moved to Berlin from Bonn.

Hungary
Poland

Countries Formed out of the Former USSR
Belarus
Georgia

Putting this former Soviet republic in Europe is a close call. Bordered on the south by Turkey and Iran, Georgia could be considered geographically part of Asia, but in culture and history it is more European. Because it was the birthplace of the Soviet dictator Joseph Stalin, some Europeans may be inclined to let Asia claim the murderous Stalin as its native son.

Ukraine
Moldova
Russia
Straddling both Europe and Asia, Russia is the largest of the former Soviet Union's fifteen republics.

The Baltic Republics
Estonia
Latvia
Lithuania
After Stalin and Hitler signed their secret nonaggression pact in 1939, Hilter gave the Soviets a free hand to move on the three Baltic Republics, which were seized by the Soviet Union in 1940. Half a century later, the three small states bordering on the Baltic Sea in northwestern Europe boldly declared their independence from the Soviet Union. Late in 1991, after the aborted coup against Mikhail Gorbachev, they were finally recognized by the Western nations as independent.

Island States
Iceland
For reasons of culture and language, this rugged little island, where most residents get their heat from geothermal sources, can also be thought of as a Scandinavian country.

Ireland
Malta
The most-bombed place in World War II, Malta is home to the most famous bird figurine in the history of detective novels, *The Maltese Falcon*.

United Kingdom
What is now called the United Kingdom, once the seat of a vast British Empire, is made up of England, Scotland, Wales, and Northern Ireland. The union took place over three centuries, beginning in 1536 when King Henry VIII merged England and Wales under a single government. Scotland was joined to England under King James I of England, the cousin of Queen Elizabeth I. In 1707, the Act of Union formally created the Kingdom of Great Britain, with a unified Parliament in Westminster. In 1801, Ireland was joined to Great Britain, but in 1921, the largest part of Ireland won independence as the Irish Free State. Six northern counties, mostly Protestant, remained part of the United Kingdom—a source of continuing strife in Northern Ireland, where the Roman Catholic minority favors unification with the rest of Ireland.

A resurgent independence movement has also taken root in Scotland, inspired by North Sea oil revenues and the nationalist fever that has spread through Europe in the past five years.

Miscellaneous Leftovers
Andorra
A small semi-independent state tucked into the Pyrenees Mountains between France and Spain.

Gibraltar
This 2.25 square mile point commands the strategic passage in and out of the Mediterranean and is a self-governing dependency of the United Kingdom. While Spain wants it back, the people of Gibraltar have voted to maintain the status quo.

Isle of Man
Another self-governing dependency of the United Kingdom located in the Irish Sea between Ireland and Great Britain.

Monaco
A tiny, hilly wedge on the French Mediterranean coast, this principality is about the size of New York's Central Park. It is the second smallest independent state in the world after the Vatican City.

San Marino
A landlocked enclave within the mountains of central Italy, this is the world's smallest republic with an area of 24 square miles (61 square km), one tenth the size of New York City. Founded by a Christian saint in the fourth century as a refuge against religious persecution, it is also the world's oldest republic.

Vatican City
The sovereign papal state located within Rome, Italy, Vatican City is the world's smallest nation.

What About Greenland and Cyprus?

Greenland, the world's largest island (if you don't count Australia), is an autonomous region of Denmark. But it has withdrawn from the European Community and geographically is considered a part of North America.

Cyprus is currently divided into Greek and Turkish sections, the Republic of Cyprus and the Turkish Republic of Northern Cyprus. Although many count it part of Asia, it belongs to the Council of Europe. Turkey itself has both European and Asian territory, but for reasons of history and culture is more readily identified with Asia.

Why Is Australia a Continent? Isn't It Just Another Island?

The Thorn Birds, Crocodile Dundee, and *Mad Max* have all helped promote Australia's romantic image in America. So put another shrimp on the "barbie" and ponder this. Is Australia the world's largest island? Or is it the world's smallest continent? Putting it simply, the answer is . . . yes.

With an area of 2,966,200 square miles (7,692,300 square km), Australia is both the world's smallest continent and its largest island. Its population of 17,500,000 is the smallest of the six inhabited continents. Five countries are larger than the continent of Australia. (In size order, they are Russia, Canada, China, the United States, and Brazil.)

The first Europeans to reach Australia were the Dutch, when Captain Abel Tasman explored the island in 1606 and called it New Holland. Captain James Cook claimed possession for the British in 1770. A few years later, when the British lost one of their other little possessions in America's War of Independence, they decided they needed a new offshore home for their convicts. Australia was then known as New South Wales, and a penal colony called Botany Bay was established near the site of present-day Sydney. The first convicts, along with a wave of British settlers looking for a new start, arrived in 1788. The prisoners continued to come for almost fifty years, more than 160,000 of them in all; the settlers never stopped coming, especially when there were major gold rushes in 1851 and 1892. The free settlers established six colonies, which later became states. In 1901 they were joined in the Commonwealth of Australia, combining the British parliamentary system with United States federal experience.

It was not until 1801 that mapmakers figured out that this floating jail was a sixth continent and not just made up of several islands. That's when it was christened Australia, in honor of the great mythological southern continent Terra Australis, which had been the subject of speculation since the time of the ancient Greeks.

NAMES: *Who Swallowed the Sandwich Islands?*

You won't find the Sandwich Islands on the maps anymore. They've been swallowed whole. That is, what once were called the Sandwich Islands have been restored to the name used by the natives of this volcanic island chain—Hawaii.

The discovery and naming of the Sandwich Islands by Captain James Cook is a single chapter in one of the most extraordinary tales of exploration and discovery in history. Born the son of a farmer, James Cook became the greatest seagoing explorer of all time. In the course of three voyages over a dozen years, he accurately filled in more details on the world map than anyone in hundreds of years. Modest, honest, and humane, he opened up the Pacific for all of Europe, for better or for worse.

When Cook set sail in 1768 on the first of his three voyages, there were already hints at what the Pacific held. From the time of Magellan and Sir Francis Drake, the first Englishman to sail around the world, stories of idyllic islands and wondrous places had filtered back to Europe. But even after the voyages of sailors like the Dutchman Tasman, who reached Australia, Europeans did not comprehend the vastness of the Pacific, the ocean which occupies more of the earth than all its land put together. Sailing in a small ship built for carrying coal, Cook set out with two sets of orders, one public, the other sealed. His first mission was to reach Tahiti (or Otaheite) in time to observe the passage—or transit—of the planet Venus across the face of the sun, an event that was predicted to occur on June 3, 1769, and then not again for another century. By observing this phenomenon from several points on the globe, British astronomers hoped to accurately calculate the earth's distance from the sun and make celestial navigation far more reliable and safe. This purely scientific objective, so different from those of previous eras of exploration when legends of gold or riches provided the sole motivation, was typical of the British approach: an Enlightenment-era attempt to expand knowledge of the world and improve navigational safety. It was this same spirit that would put naturalists on British navy ships, including, a few years later, a young man named Charles Darwin.

Cook's second, secret orders were more mercenary. He was to search for the legendary southern continent, which the British hoped would provide another new world to colonize, since one of their existing colonies was beginning to show signs of a rebellious adolescence by the mid-1700s.

Tahiti was already known to Europe because the island had been visited by a British ship in 1766 and by the French explorer Louis Antoine de Bougainville (1729–1811). Both brought back reports of innocent and carefree natives living in a state that sounded to most Europeans as if paradise had indeed been discovered. Tales of unashamed native girls and women willing to have sex freely and openly, often in exchange for a tenpenny nail, were soon the talk of Europe. One story from Bougainville's account of his voyage around the world was typical of what Europe was hearing. After reaching Tahiti, Bougainville's ship was surrounded by canoes of naked women. Marines were called out to keep order among the sailors. But the ship's cook could not be restrained. He made it to the beach, where he was immediately dragged into the bushes by a band of women, stripped, and publicly performed "the act for which he had come ashore." After returning, the cook reportedly told his captain that whatever his punishment for disobeying orders, it would not be more terrifying than the ladies on the beach. (Bougainville also brought back samples of plant life he had found on his voyage, including the genus which is named for him, the woody flowering shrub bougainvillea.)

It was in this strange new world that Cook's crew arrived after the long, difficult passage. After observing the transit of Venus, Cook went on to discover the Society Islands, then sailed around New Zealand and found it was not part of the mysterious southern continent. He explored and charted more than two thousand miles of the eastern coast of New Holland (Australia), stopping and naming Botany Bay and discovering (and almost being shipwrecked on) the Great Coral Reef.

On his second voyage of some seventy thousand miles between the years 1772 and 1775, Cook continued to seek the southern continent, now outfitted with Harrison's chronometer, a device that accurately measured longitude. He neared Antarctica twice, finally eliminating the romantic notions of a southern continent, explored New Zealand and the Hebrides Islands, and by experimenting with diet, eliminated scurvy among his men. This vitamin-deficiency disease had taken the lives of many sailors. But by the simple means of including citrus fruits and sauerkraut in their diet, Cook was able to keep his sailors from suffering. Despite the hardships of sailing in the frozen waters near the Antarctic Circle, where icebergs towered over the ships like mountains, Cook lost only one man.

On his final voyage, Cook was assigned to tackle one more legend, the existence of a northwest passage connecting the Atlantic and Pacific Oceans through the top of North America, a geographic notion that had inspired nations to send out sailors since the European discovery of America. On this voyage, Cook reached the Hawaiian Islands, naming them the Sandwich Islands after the Earl of Sandwich (the man who ate his meat between two pieces of bread), and then continued on to the western coast of North America. For several months, he carefully mapped the coast from Oregon north to Alaska, sailing almost as close to the North Pole as he had to the South Pole on his second voyage. Although he did not find the passage, he had hopes of returning to search for it again and he headed southwestward to pass the winter in the Sandwich Islands. But while there, Cook's boats were constantly visited by natives who stole bits and pieces of metal from them. When a small boat was taken by some islanders, Cook went ashore to recover it and scuffled with the natives. He was stabbed, then drowned and dismembered.

Cook's Sandwich Islands are today the islands of Hawaii, America's fiftieth state. Located about twenty-four hundred miles west-southwest of San Francisco, California, Hawaii is a chain of some one hundred thirty volcanic islands and islets stretching over fifteen hundred miles in the North Pacific. The state is centered on eight main islands—Hawaii, Kahoolawe, Maui, Lanai, Molokai, Oahu, Kauai, and Niihau. First settled by Polynesians sailing from other Pacific islands between AD 300 and 600, Hawaii was untouched by Europe until Cook's arrival in 1788.

The first Americans to arrive there were missionaries who began the process of westernization in 1820. A largely ignored native kingdom during the nineteenth century, Hawaii became more valuable to American interests when U.S. naval power was expanded in the late part of that century. Treaties in 1875 and 1887 gave the United States the rights to a naval way station established at Pearl Harbor on Oahu. Commercial exploitation soon followed as American interests moved in on Hawaii's sugar business and then introduced the pineapple in

1898. These American businesses eventually spearheaded a revolt that deposed the Hawaiian queen in 1893, establishing a republic under the first president, Sanford B. Dole.

The United States later annexed Hawaii as part of its expansion in the Pacific that included the acquisition of the Philippines following the Spanish-American War. The nation's leadership had begun an aggressive plan to build American naval power in the Pacific, prompted by the influential book *The Influence of Sea Power Upon History, 1660–1783* by Captain Alfred Mahan, and his enthusiastic disciples Massachusetts senator Henry Cabot Lodge and Theodore Roosevelt (first as a navy undersecretary and later as president). Pearl Harbor was established as a U.S. naval base in 1908 but was not in the American consciousness until it was attacked by the Japanese on December 7, 1941, bringing the United States into war against Japan and its ally, Germany. After the war, Hawaiians unhappy with their lack of representation lobbied for statehood, which was granted on August 21, 1959.

Geographic Voices   From the journal of Captain James Cook

The young girls, whenever they can collect eight or ten together, dance a very indecent dance, which they call *Timorodee*, singing most indecent songs and using most indecent actions, in the practice of which they are brought up from earliest childhood. In doing this they keep time to a great nicety. This exercise is, however, generally left off as soon as they arrive at years of maturity, for as soon as they have formed a connection with man they are expected to leave off dancing *Timorodee*.

One amusement or custom I must mention, though I must confess I do not expect to be believed, as it is founded upon a custom so inhuman and contrary to the principles of human nature. It is this: that more than one half of the better sort of inhabitants have entered into a resolution of enjoying free liberty in love, without being troubled or disturbed by its consequences. These mix and cohabit together with the utmost freedom, and the children who are so unfortunate as to be thus begot are smothered at the moment of their birth. Many of these people contract intimacies and live together as man and wife for years, in the course of which the children that are born are destroyed. They are so far from concealing it that they rather look upon it as a branch of freedom upon which they value themselves. They are called *Arreoys*, and have meetings among themselves where the men amuse themselves with wrestling, etc., and the women in dancing the indecent dance before-mentioned, in the course of which they give full liberty to their desires.

During the first voyage, Cook also encountered the aborigines of Australia and wrote about them.

From what I have said of the natives of New Holland [Australia] they may appear to some to be the most wretched people upon earth; but in reality they are far more happier than we Europeans, being wholly unacquainted not only with the superfluous but the necessary conveniences so much sought after in Europe; they are happy in not knowing the use of them. They live in a tranquility which is not disturbed by the inequality of conditions. The earth and sea of their own accord furnishes them with all things necessary for life.

What Is the International Date Line?

Time, as we think of it today, is a relatively recent invention. The internationally accepted "standard" time was established a little more than a hundred years ago. Before then, people in different places set their clocks—another fairly recent invention—to arbitrary notions of the hour, which usually came from when the sun entered their area. There was a time, for instance, when clocks in Camden, New Jersey, were set differently from those in nearby Philadelphia.

As the world entered the modern scientific age, the advent of oceanic navigation, steamship timetables, telegraphic communication, and train schedules all demanded coordination; the world needed to get on one time standard. In 1883, they did it and cleverly called it standard time.

Meeting in Washington, DC, the time setters divided the world into twenty-four zones of one hour each, the time it takes the sun to cross each zone. These zones were located 15° of longitude apart (360° divided by 24 hours equals 15°). Because a starting point was needed, Greenwich (near London), site of the most prominent astronomical observatory of its day, was selected as 0°—the prime meridian. Lines of longitude were then counted either east or west from Greenwich. Since the sun rises in the east, the day began there. At any given hour, standard time is later in the day in points to the east, earlier in points to the west.

In practical terms, that means when it is 5 P.M. in London, it is 10 P.M. in Karachi, Pakistan, five time zones to the east. At the same moment, in Kuala Lumpur, capital of Malaysia, and Manila in the Philippines, it is 1 A.M. and the next day has begun. In Tokyo, nine time zones earlier, it is 2 A.M. and in Melbourne, Australia, it is 3 A.M. the next day.

At points west of London, in South America and North America, it is earlier in the day. At 5 P.M. London time, it is 2 P.M. in Rio de Janeiro; 12 noon in New York City and Quebec; 9 A.M. in San Francisco and Vancouver, Canada. Way out in Anchorage, Alaska, it is 8 A.M.

While this gave the world a uniform clock, it raised another question: where does one day turn into another? The logic behind this question is simple. Twelve time zones to the west from London, it is twelve hours earlier. Twelve time zones to the east, it is twelve hours later. One place can't be both.

For example, if it is 5 P.M. on Sunday in London, it is twelve hours later to the east, or 5 A.M. on Monday. Yet at the same time, it is twelve hours earlier to the west, or 5 A.M., on Sunday. But how can it be two different days in the same place? The simple solution was to establish another of geography's imaginary lines on the meridian of 180°, directly opposite the prime meridian at Greenwich. This is where East literally meets West. In 1883 this line became known as the international date line, the point at which the calendar day changes by one day as it is crossed. Fortunately, this line happens to be in the middle of the Pacific Ocean (for the most part), where it can cause the least confusion. The date line does zigzag its way around several spots to keep some places within the same time zone.

In practical terms, the date is one day earlier on the eastern side of the line; and it is one day later on the western side of the line. A traveler crossing the date line westward advances the calendar—for example, 5 A.M. Sunday becomes 5 A.M. Monday. A traveler crossing the date line eastward has to put back the calendar from 5 A.M. Monday to 5 A.M. Sunday. Depending on which side of the date line the traveler is on, he is now twelve hours different from London, either earlier or later.

What Does the Continental Divide Divide?

Let's start by explaining what it doesn't divide. A continental divide does not cut continents into equal pieces. One of those invisible boundaries that geographers find so useful, a continental divide is a line of high mountain peaks marking the point where a continent's rivers begin to flow in opposite directions. As the water flows down one side of the range or the other, its ultimate destination is set. In North America, it is a ridge of high ground running irregularly through the Rocky Mountains and Mexico's Sierra Madre, and separating eastward- from westward-flowing streams. The waters that flow eastward empty into the Atlantic Ocean, chiefly by way of the Gulf of Mexico; those that flow westward primarily empty into the Pacific, although some drain into the deserts of the Southwest and never reach the ocean.

Each of the continents has a similar mountain range or high ridge that directs the flow of rivers in opposite directions. In South America, the continental divide follows the course of the Andes, and rivers flow either to the Pacific or, like the Amazon, into the Atlantic. Europe's divide separates those streams that drain into the Atlantic and Arctic Oceans from those that flow into the Mediterranean and Black seas. In Asia, the divide separates rivers that flow into the Indian Ocean, including the Ganges and Indus, from those that empty into the Arctic and Pacific Oceans. Africa's divide separates streams that drain into the Indian Ocean on the east and the Atlantic on the west. And in Australia, the continental divide separates Pacific-draining rivers from waters that empty into the Indian Ocean.

Geographic Voices   Meriwether Lewis, from *The Journals of Lewis and Clark*

11th. February Monday 1805.

About five Oclock this evening one of the wives of Charbono was delivered of a fine boy. it is worthy of remark that this was the first child which this woman had boarn, and as is common in such cases her labour was tedious and the pain violent; Mr. Jessome informed me that he had frequently administered a small portion of the rattle of the rattle-snake, which he assured me had never failed to produce the desired effect, that of hastening the birth of the child; having the rattle of a snake by me I gave it to him and he administered two rings of it to the woman broken in small pieces with the fingers and added to a small quantity of water. Whether this medicine was truly the cause or not I shall not undertake to determine, but I was informed that she had not taken it more than ten minutes before she brought forth perhaps this remedy may be worthy of future experiments, but I must confess that I want faith as to it's efficacy.

The expedition of Meriwether Lewis (1774–1809) and William Clark (1770–1838) across a largely unknown continent is one of America's great true adventure stories. Although President Thomas Jefferson had been secretly planning such an expedition for some time in order to secure trading rights in this territory and prepare a defense against British attempts to take the land, the purchase of the Louisiana Territory from Napoleon in 1803 gave it new urgency and legitimacy. Lewis and Clark had been given ambitious directions by Jefferson. They were to attempt to find a waterway clear across the continent to the Pacific Ocean, to map and catalog the plants and animals of the unknown American West, and to study the customs of native Indians while they were at it.

More than curiosity was at stake. The British had their own claims to America's Northwest. For reasons of defense and commerce, Jefferson wanted to know exactly what the new territory meant to America. With this purchase, Jefferson had doubled the size of the country. A great intellect and a brilliant politician, Jefferson also had a practical side: he knew that there was great trading and commercial potential in this new land, and he wanted to assure that Americans, not the British, French, Spanish, or Russians, would profit from the land.

The new mother described by Lewis was Sacagawea, a teenaged Shoshoni Indian who had been captured by another tribe five years earlier and then either bought or won by Toussaint Charbonneau, a French trapper. Lewis and Clark took on Charbonneau as a guide specifically because of Sacagawea's value as an interpreter among the Indian tribes that the explorers anticipated meeting further west.

A valued member of the expedition, all the time carrying her newborn son on her back, Sacagawea died in 1812 at age twenty-three. Her son, Jean Baptiste, nicknamed Pomp, was raised by Clark and later traveled to Germany with a European prince. Pomp eventually returned to America and became a trapper and guide.

Lewis and Clark completed this remarkable journey with the loss of only a single man, to appendicitis. And they did it by establishing peaceful relations with the Indians, with the exception of one brief skirmish after some horses were stolen. It is a tragic pity that Jefferson's humanistic orders for dealing with the Indians, and Lewis and Clark's success in carrying out Jefferson's peaceful directives, did not set the standard for future relations between the federal government and the Indians.

When Lewis and Clark's description of the natural wonders and riches they had seen was made public, it helped fuel an empire-building spirit that was eventually called Manifest Destiny and which changed the course of American and world history.

What's So Bad About the Badlands?

Lewis and Clark's travels up the Missouri River led them through the extraordinary heartland of America. Beginning in St. Louis, the point at which the Missouri joins the Mississippi, they traveled America's longest river west across what is now the state of Missouri. They rowed upriver north through the Great Plains of Kansas and Nebraska, on into the Dakotas, finally stopping to winter near the site of North Dakota's present capital of Bismarck. In the spring, they set out again, the Missouri taking them west toward the majestic Rockies. As they neared the junction of the Missouri and the Yellowstone Rivers near today's Montana–North Dakota border, they passed through what is today North Dakota's Badlands.

Talk about giving a place a bad name. You've heard of those places that are nice to visit but you wouldn't want to live there? The Badlands might be what they had in mind when they coined that expression. The idea of calling this territory "bad lands" dates at least to the Sioux, who called them *mako sica*, literally "bad land." But for thousands of years, Native Americans had discovered good uses for the treacherous, forbidding landscape of the Badlands. Instead of trying to kill bison with crude weapons, they stampeded great herds of the animals over their stark, steep cliffs. French fur traders, the first Europeans to penetrate this north-central section of America, agreed with the Sioux assessment and called the region *mauvaises terres à traverser* ("bad lands to travel across").

Once a flatland beneath an ancient inland sea, the Badlands are stark, arid regions, seamed and lined with deep gullies that have been cut by occasional heavy rains, often accompanied by violent thunderstorms. The unequal resistance of rocks—softer rock erodes more readily than harder rock—leaves tall columns and platforms of stone standing out above the surrounding land. The Badlands don't generally receive enough regular rain to support a covering of grass or other vegetation. When the sudden rains do come, they suddenly turn the landscape into a gluelike mud. Although some grasses survive in this inhospitable climate—115° in summer, 30° in winter—the Badlands are almost valueless for agriculture or pasture land. The farmers who were given some of this land, which had been taken from the Indians in the late nineteenth century, quickly learned that.

In 1876, after the Indians' victory at the Battle of Little Big Horn, the Sioux were placed on a reservation near the biggest and baddest part of the region, in western South Dakota. Now the Badlands National Park, it is 243,302 acres of ravines and sharp ridges of multicolored shale. Frustrated and starving, the young Sioux began a religious revival called Ghost Dancing around 1889. A reaction against white ways that called for a return to Indian traditions, the movement turned bellicose and a dangerous element was added when the Ghost Dancers were told that their magic shirts would protect them from federal bullets. A U.S. Army crackdown on the Indians began. One band of Sioux, led by a sick and aging Chief Big Foot, came out of the Badlands and surrendered to the army. But as the Indians were disarmed, a shot was fired and the result was an outright massacre of the Indians, most of them old, or women and children. The site of this massacre was Wounded Knee Creek in the Badlands.

The other great stretch of American badlands is that seen by Lewis and Clark in western North Dakota, now the site of Theodore Roosevelt National Park. As a young man, Roosevelt was among the first easterners who set out for the West lured by romantic tales of adventure and great hunting. Although he came as a hunter, Roosevelt quickly came to appreciate the value of the preservation of land and animals and is rightly considered one of the founders of the American conservation movement, one reason he is the only president to be honored with a national park.

Both the Badlands and Roosevelt national parks attract substantial tourism, but in the past they have held even greater appeal for paleontologists. Fossils have been found in the Badlands from as far back as 80 million years, when the area was an ocean bed, and include a fossil turtle twelve feet long.

Although the two Badlands of the Dakotas are the most famous, the term *badlands* is also applied to similar regions in Asia, such as in parts of Mongolia's Gobi Desert.

What Is a Butte?

First of all, say it like this: *beaut* as in *beautiful,* not like the body part you're sitting on.

This is the geographic word for what Bart Simpson's haircut looks like. Typical of the stark scenery of the Badlands, a butte is a hill that rises sharply from the surrounding area and has sloping sides and a flat top. A characteristic formation of the plateau region of the western United States, these flat-topped hills are formed when hard rock sits on top of weaker rock like a helmet, keeping the weaker rock beneath from being worn down by natural forces of erosion.

Sounds just like a mesa? Well, it is. Except that the butte (from the French word for a "mound behind targets") is smaller than a mesa, which is the Spanish word for "table." Buttes are often produced from mesas that have been reduced in size through erosion. Going back a geological step, mesas are eroded forms of plateaus, which are large highland plains raised above the surrounding land.

The word *mesa* is one of the many remnants of the Spanish domination of the New World for almost one hundred years before the English arrived at their first permanent settlement, Jamestown, Virginia. There is a rich array of Spanish terms that have become part of the language of America's geography. Obvious examples can be found in the many Spanish-named places of South America, Mexico, and the American West and South. Rio Grande, Los Angeles, San Diego, Ecuador, Florida ("feast of flowers"), Colorado, Montana, and Sierra Madre are just a few. Besides *mesa*, the Spanish left behind such geographic terms as *cañon* (canyon in English); *arroyo* (for a deep gully cut in the desert by an intermittent stream; the Arabs call it a *wadi*; in India it is a *nullah*); the *chaparral*, an area of low, dense scrub-brush. These go along with such typical southwestern terms as bronco, corral, lasso, ranch, and rodeo, as Spanish words that moved from the Spanish-influenced dialect of the Southwest into modern American English.

Geographic Voices   Mark Twain's description of the Continental Divide from *Roughing It* (1872)

We bowled along cheerily, and presently, at the very summit, we came to a spring which spent its water through two outlets and sent it in opposite directions. The conductor said one of those streams which we were looking at, was just starting on a journey westward to the Gulf of California and the Pacific Ocean, through hundreds and even thousands of miles of desert solitudes. He said that the other was just leaving its home among the snow-peaks on a similar journey eastward—and we knew that long after we should have forgotten the simple rivulet it would still be plodding its patient way down the mountain sides, and canyon-beds, and between the banks of the Yellowstone; and by and by would join the broad Missouri and flow through the unknown plains and deserts and unvisited wildernesses; and add a long and troubled pilgrimage among snags and wrecks and sand-bars; and enter the Mississippi, touch the wharves of St. Louis and still drift on, traversing shoals and rocky channels, then endless chains of bottomless and ample bends, walled with unbroken forests, then mysterious byways and secret passages among woody islands, then the chained bends again, bordered with wide levels of shining sugar-cane in place of the somber forests; then by New Orleans and still other chains of bends—and finally, after two long months of daily and nightly harassment, excitement, enjoyment, adventure, and awful peril of parched throats, pumps and evaporation, pass the Gulf and enter into its rest upon the bosom of the tropic sea, never to look upon its snow-peaks again or regret them.

I freighted a leaf with a mental message for the friends at home, and dropped it in the stream. But I put no stamp on it and it was held for postage somewhere.

Before Samuel Langhorne Clemens (1835–1910) first signed his name "Mark Twain" and became famous for his tales of the Mississippi River, *Tom Sawyer* and *Huckleberry Finn*, he spent several years in the American West. In 1861, after two weeks as a soldier in the Confederate militia, Clemens joined his older brother Orion, who had been appointed secretary of the Nevada Territory, on a stagecoach excursion to the West. Expecting to spend three months there, Clemens spent the next five years as a prospector, miner, and journalist. From Nevada, he went to San Francisco and then to the Sandwich Islands as a correspondent.

In time, under the pseudonym Mark Twain, he became one of the most popular humorists and correspondents in America. But before he wrote the novels most people associate with him, he was a successful travel writer. One of his earliest books, *The Innocents Abroad* (1869), was an account of his voyage to Europe and the Holy Land and was a great financial success. In 1872, he published *Roughing It*, his equally popular book about his years in the West. These light-hearted accounts of stagecoach travel, Indians, frontier society, the peculiarities of the Mormons, and the customs of the West—tempered by Twain's cutting wit, hilarious imagination, and jaundiced eye—are still fresh more than a hundred years later.

Milestones in Geography III
1600–1810

c. 1600 The beginning of the scientific revolution in Europe. Among the significant personalities are German astronomer Johannes Kepler (1571–1630), English philosopher-scientist Francis Bacon (1561–1626), Italian physicist-astronomer Galileo Galilei (1564–1642), and French mathematician-philosopher René Descartes (1596–1650).

1602 The Dutch East India Company is founded, the first modern "public" company, for the purpose of expanding trade in Asia. In 1609, the company began shipping tea to Europe from China. In 1619, the company established a colony in Batavia (modern Jakarta, Indonesia), marking the beginnings of the Dutch Empire in the East Indies.

1607 The first permanent English settlement in America is established at Jamestown, Virginia.

1608 Under explorer Samuel de Champlain (1567–1635), the French establish a colony at present-day Quebec. Champlain finds and explores the Great Lakes and opens up Canada to the fur trade, which will bring a generation of French explorers and priests to North America.

1609 The telescope is invented in Holland by Hans Lippershey. Italian physicist-astronomer Galileo Galilei sets out to improve on it and, during the next few months, makes some of the most important discoveries in the history of astronomy. His observations include the discovery that the moon's light is reflected sunlight; that the moon's surface is covered by craters and mountains; and that the Milky Way is composed of separate stars. He also proposed the existence of "sunspots." His 1613 treatise on sunspots supported the Copernican theory that the earth revolved around the sun. But the Roman Catholic Church forced him to stop teaching this new doctrine, and under the threat of torture, Galileo recanted. He was allowed to return to Florence and then went blind a few years before his death.

1620 The Puritan settlers known as Pilgrims, sailing on the *Mayflower*, land in New England and establish the Massachusetts Bay Colony.

1640s Jan Blaeu's *Atlas Major* is published in twelve volumes. By this time the Dutch East India Company has monopolized trade with the East and created a huge seagoing empire based in Amsterdam. Although the Dutch attempted to keep much of the information gained by Dutch captains secret, a trade in maps and atlases soon boomed. Most of what Dutch sailors had discovered in sailing to the East was incorporated into Blaeu's collection and was soon available throughout Europe.

1645 Dutch navigator Abel Tasman (circa 1603–59) circumnavigates Australia and discovers New Zealand.

1652 Cape Colony in South Africa is founded by the Dutch.

1665 At age twenty-three, Isaac Newton (1642–1727) works out the general principles of his universal laws of gravitation.

1668 The British East India Company, founded in 1600, obtains control of Bombay and eventually controls all of India and the Himalayan areas, as well as dominating trade with China. A company that was a government unto itself, the British East India practically ruled these places until well into the nineteenth century, when the British Crown took possession and they became Imperial colonies.

1670 French explorer René-Robert Cavelier de La Salle (1643–87) descends the Ohio River thinking it will flow into the Pacific. He reaches the Mississippi River instead and over the course of two hard years follows it to the sea. Claiming the vast territory of the Mississippi basin for France, he names it Louisiana in honor of the French king. In 1684, he attempted to find the Mississippi from the Gulf of Mexico but was unable to locate the mouth of the river. After two years of searching in vain, La Salle's men mutinied and murdered him.

1675 Greenwich Observatory, intended to improve navigation by providing more accurate information about the position of stars and planets, is founded and becomes the world's leading scientific center.

1687 *Principia Mathematica* by Sir Isaac Newton is published. In this masterwork, Newton sought to explain all physical phenomena by a few generalized laws.

1696 Completion of *Planisphere Terrestre*, one of the first scientifically compiled world maps, at the Paris Observatory.

1698–99 English astronomer Edmund Halley (1656–1742) maps magnetic variation in the Atlantic Ocean, significant because of the effect of these variations on ships' compasses. Halley also predicted the regular return of the comet that has been named after him.

1736–44 The French Academy mounts ambitious expeditions to Lapland and Peru in an attempt to prove that the earth is flattened at the poles, as predicted by Newton. Both parties suffer extreme hardships. The Lapland expedition was first to accomplish the feat; the expedition to Peru took nearly ten years and returned to learn that the other expedition had successfully accomplished its mission many years earlier.

1762 English clockmaker John Harrison's No. 4 Marine Chronometer accurately keeps time at sea, ending a long scientific treasure hunt. A clock that is unaffected by temperature, ship's motion, or changes in gravitational forces, it was a landmark in the history of navigation, allowing sailors to precisely determine their longitude, instead of depending on the unreliable and unsafe method of dead reckoning.

1764 A survey of Pennsylvania and Maryland is begun by British surveyors Charles Mason and Jeremiah Dixon. It eventually produces the Mason-Dixon line, which later effectively divides northern and southern states.

1768–71 Captain Cook makes the first of three voyages to the Pacific. The second took place between 1772 and 1775. The third, during which Cook was killed in Hawaii, lasted from 1777 to 1779.

1782 Working for the British East India Company, James Rennell produces the first edition of his *Map of Hindoostan*, the first scientifically accurate map of India.

1787 Triangulation across the English Channel links surveys of Britain and France.

1791 Beginning of the British Ordnance Survey, the British government's official arm for mapping the entire country.

1792–94 British explorer George Vancouver (1758–98) surveys and maps the northwestern coast of North America. He sails a hundred miles up the Columbia River (as far as the site of present-day Portland, Oregon). The Columbia was named earlier by American captain Robert Gray. These two expeditions established competing British and American claims to the territory that went unsettled until the mid-nineteenth century.

1792–93 Scottish explorer Alexander Mackenzie completes a transcontinental passage across Canada, becoming the first European to cross the Rockies to the Pacific.

1793 The French *Carte de Cassini*, the first scientifically conducted national survey, is completed.

1802 The Great Trigonometrical Survey of India is begun, a project that will last for nearly a hundred years, culminating in the mapping of the Himalayas. In August 1913, British and

Russian survey parties meet in Kashmir to link their separate surveys of Central Asia.

1803 The Louisiana Purchase. For $15 million, about two cents an acre, President Thomas Jefferson more than doubles the size of the United States by purchasing France's holdings in North America from Napoleon. The French emperor had been frustrated in an attempt to retake Haiti from the former slaves who had revolted and set up a republic there. Unsuccessful against the former slaves, the French army sent to retake the island also succumbed to yellow fever, frustrating Napoleon's designs on North America. Short on cash with which to continue fighting in Europe, Napoleon agreed to the sale of the massive French territory.

1804–6 Lewis and Clark make an expedition to find out if the Missouri River can be traveled all the way to the Pacific and to survey the lands acquired in the Louisiana Purchase.

1810 William Clark's map of the American West is published.

3

If People Were Dolphins, the Planet Would Be Called Ocean

Alone, alone, all, all, alone;
Alone on a wide, wide sea.

—Samuel Taylor Coleridge
The Rime of the Ancient Mariner

There is, one knows not what sweet mystery about this sea, whose gently awful stirrings seem to speak of some hidden soul beneath.

—Herman Melville
Moby-Dick

And then, as never on land, he knows the truth that his world is a water world, a planet dominated by its covering mantle of ocean, in which the continents are but transient intrusions of land above the surface of the all-encircling sea.

—Rachel Carson
The Sea Around Us

How Many Oceans Are There?
What Is the Difference Between an Ocean and a Sea?

We eat seafood, but take an ocean cruise on ocean liners. Scientists study oceanography rather than "seaography." Sea horses live in the ocean. In the summer, people go to the seashore, where they rent an oceanfront house. When you feel lost, you're "at sea" instead of "at ocean." And in the song "America the Beautiful," the lyrics don't say from "ocean to shining ocean." Confused? No wonder.

The ocean has a special allure. People are drawn to it and find tranquility in the gentle lapping of waves on a beach or something awesome in the majestic crashing of the big breakers. The endless cycle of tides and waves hints powerfully at the earth's timelessness.

The ocean is our past. Life began in the oceans. The ocean is the home of the greatest number of living things on this planet. The ocean even seems to be in our blood. Scientists tell us that the chemical makeup of human body fluids is remarkably similar to that of ocean water.

And the ocean is our future. But it is increasingly a suspect future. What once was thought to be too big to be contaminated is showing the wear and tear of the dumping of sewage, garbage, and industrial waste; accidental oil spills; and even an act of war that unleashed millions of barrels of oil into the Persian Gulf's sensitive environment. Equally alarming is the growing threat that an upward shift in global temperatures will raise ocean levels in the near future, with devastating consequences for coastal areas.

People who have sailed across the oceans or flown above them can begin to grasp the vastness of the earth's great waters. But it probably wasn't until the space missions of the 1960s began to send back snapshots of the planet that people realized how wet and blue our planet truly is. Viewed from aloft, there isn't a lot of earth on Earth. Those picture postcards from space didn't show great lands separating many oceans and seas, but One Great Ocean, broken occasionally by parcels of land. As those space pictures made evident, the oceans are the earth's most prominent feature. They dictate our weather and have determined human history.

The One Great Ocean is an interconnecting body of saltwater covering almost three fourths of the planet, more than 142 million square miles. That is twice the surface area of Mars and nine times the surface area of the moon. Of the earth's remaining 30 percent, 24 percent is untillable desert, tundra, glacial ice, and mountaintops, leaving humans about 6 percent on which to farm. All of which makes you feel a little like the proverbial drop in a bucket.

How Many Oceans Are There?

This is like the trick questions you got in high school from the "hard" teacher. In a sense, there are two answers, both of them right. Strictly speaking, there is only one ocean, the great sheet of salt water that altogether covers about 72 percent of the earth's surface and surrounds the planet's great land masses. But in more familiar terms, the One Great Ocean is divided into four principal parts, each of them known as an ocean.

The Pacific covers about 70 million square miles (181,300,000 square km) and is by far the largest ocean, containing about 46 percent of the earth's water. Bounded by the Americas on the east and Asia on its west, the Pacific is flanked by high mountain chains and is remarkable for its many small islands, many of them volcanic. The Pacific is larger than all the land in the world put together. It extends almost from pole to pole but is largely a tropical ocean, and half of the equator's length of 24,000 miles (38,500 km) lies within the Pacific. The Challenger Deep in the Mariana Trench, which extends from southeast of Guam to northwest of the Mariana Islands in the Pacific Ocean, is the deepest place in the oceans of the world, plunging to 36,198 feet (11,040 meters); if Mount Everest were dropped into the Mariana Trench, it would not break the surface of the Pacific.

The Atlantic is the second largest ocean, containing about 23 percent of the world's water. Although its extent from north to south is about the same as the Pacific's, it is much narrower and only half the size of the Pacific, covering about 32 million square miles (82,217,000 square km). An S-shaped body bounded by the Americas on the west and Europe and Africa on the east, it is not as deep as the Pacific and contains many fewer islands. Lying between highly industrialized continents, the North Atlantic carries the greatest proportion of the world's shipping. Although half of the world's fish are caught in the Atlantic, with much of that catch coming from the Grand Banks, an underwater plateau off Newfoundland, pollution and overfishing by commercial fleets threaten the existence of a number of Atlantic fish species.

The Indian, called the Erythraean Sea in ancient times, is the third-largest ocean. Slightly smaller than the Atlantic, it covers about 28 million square miles (73,426,500 square km), and holds 20 percent of the world's water. Bounded by Asia to the north, Antarctica to the south, Africa to the west, and Australia and Indonesia to the east, the Indian Ocean is divided in two by India, forming the Arabian Sea on one side and the Bay of Bengal on the other. Ninety percent of the Indian Ocean lies south of the equator.

The Arctic, lying within the Arctic Circle and surrounding the North Pole, is the smallest ocean, with an area of about 5.5 million square miles (13,986,000 square km) containing 4 percent of the world's water. Connected to the Pacific by the Bering Strait (between Alaska and Russia) and to the Atlantic by the Greenland Sea, the Arctic is frozen year-round except at its outer margins, but recent research suggests that the Arctic is experiencing a period of warming.

Since 2000, the Great Southern, or Antarctic Ocean, which circles Antarctica, has been designated an ocean by the International Hydrographic Organization, but as noted earlier, this extension of the southern portions of the Pacific, Atlantic, and Indian Oceans is still not widely considered a fifth ocean.

Geographic Voices   Charles Darwin, from *The Voyage of the Beagle* (1839)

> The day was glowing hot, and the scrambling over the rough surface and through the intricate thickets, was very fatiguing; but I was well repaid by the strange Cyclopean scene. As I was walking along I met two large tortoises, each of which must have weighed at least two hundred pounds: one was eating a piece of cactus, and as I approached, it stared at me and slowly stalked away; the other gave a deep hiss, and drew in its head.

These huge reptiles, surrounded by the black lava, the leafless shrubs, and large cacti, seemed to my fancy like some antediluvian animals. The few dull-colored birds cared no more for me, than they did for the great tortoises.

On the 19th of August we finally left the shores of Brazil. I thank God, I shall never again visit a slave-country. To this day, if I hear a distant scream, it recalls with painful vividness my feelings, when passing a house near Pernambuco, I heard the most pitiable moans, and could not but suspect that some poor slave was being tortured, yet knew that I was as powerless as a child even to remonstrate. I suspected that these moans were from a tortured slave for I was told that this was the case in another instance. Near Rio de Janeiro I lived opposite to an old lady, who kept screws to crush the fingers of her female slaves. I have staid in a house where a young household mulatto, daily and hourly, was reviled, beaten, and persecuted enough to break the spirit of the lowest animal. I have seen a little boy, six or seven years old, struck thrice with a horse-whip (before I could interfere) on his naked head, for having handed me a mere glass of water not quite clean; I saw his father tremble at a mere glance from his master's eye.

Galápago is the Spanish word for tortoise. The first of these excerpts explains how these famed islands got their name. An island province of Ecuador, the Galápagos Islands (officially known as the Archipiélago de Colón) are a combined 3,075-square-mile island group located in the Pacific about 600 miles (970 km) west of mainland South America. The group includes six main islands, one smaller island with an airport, and eleven uninhabited islands.

Most people's image of Charles Darwin (1809–1882) is that of a somber and stern bearded Victorian man permanently linked with that of a monkey, an oversimplification of his ideas of evolution. But as these excerpts show, Darwin had a human and humane side. His account of a five-year voyage on board the surveying ship HMS *Beagle* was a great popular success in England and made Darwin a bit of a literary celebrity before he became more notorious as the father of modern evolutionary theory. The unique animal life of the Galápagos Islands, cut off from contact with other species, inspired much of Charles Darwin's revolutionary thinking. Although he had begun to formulate his theories of natural selection during the voyage, he was reluctant to publish them. Only when he received a manuscript written by a friend and fellow scientist, Alfred Russel Wallace, outlining a set of ideas remarkably similar to his own, did Darwin publish *On the Origin of Species by Means of Natural Selection* (1859).

Simple enough for the lay reader, the book was an instant success, with the first printing selling out in a single day. It also marked the opening of an enormous controversy. Darwin's idea that species gradually evolve from earlier species didn't simply challenge existing scientific notions. Overnight, his theories called into question the whole of Christian orthodoxy and the very truth of the Bible in a time when such ideas were heresy, and cause for ridicule and social disgrace. Darwin provoked even greater public outrage in 1871 with *The Descent of Man*, which put forward the idea that man and the anthropoid apes were descended from a common ancestor, now a basic article of scientific understanding.

What Is the Difference Between an Ocean and a Sea? And Are There Only "Seven" Seas?

This is another trick question. If you agree that there is only one ocean, then the seas are part of it. But for the sake of finding your way around the world, the One Great Ocean has been split up. Four oceans; many more seas. Basically, a sea refers either to a smaller division of the oceans or to a large saltwater body partially enclosed by land.

Just in case oceans and seas didn't sufficiently confuse you, there are also *bays* and *gulfs* to further complicate matters. A bay is simply a large indentation into the land formed by the sea. Much larger than bays are gulfs; large, deep inlets of the ocean or sea surrounded by land or an extensive inlet penetrating far into the land. There are some bays and gulfs that are larger than seas.

Despite the proverbial "seven seas," there are many more seas in the world. "Seven seas" is actually a colloquial expression relating not to the seas but to the oceans: the Arctic, Indian, North Pacific, South Pacific, North Atlantic, South Atlantic, and Antarctic, which isn't even an ocean. So while the expression may be familiar, it is far from geographically accurate.

The World's Principal Seas

Estimates of the size of the seas vary widely because seas don't have clearly defined boundaries. The following list of the world's major seas is in approximate size order.

South China Sea An arm of the Pacific covering more than 1 million square miles, it is a tropical sea, subject to frequent typhoons.

Caribbean Sea Named after the Carib Indians, a tribe Columbus discovered when he arrived in 1492, the Caribbean is an arm of the Atlantic. From the tip of Cuba, lying some ninety miles from Florida, to the island of Trinidad, just off the coast of Venezuela, the islands of the Caribbean Sea stretch out like a necklace whose jewels have been the source of war, conquest, and exploitation since Columbus reached these waters five hundred years ago. The vestiges of the Caribbean's colonial past remain in the continued possession of several of these islands by other countries. The Gulf Stream—the warming ocean current that influences climate on both sides of the Atlantic—originates here.

Mediterranean Sea The world's largest inland sea, lying between Africa, Europe, and Asia, it is connected to the Atlantic by the narrow Strait of Gibraltar and to the Indian Ocean by the Red Sea via the Suez Canal. Its name, meaning "middle of the earth," reflects its central importance to early history and the development of civilizations on its shores, from Egypt and the Phoenicians to the Greeks, Romans, and Arabs.

Bering Sea An extension of the Pacific, the Bering lies between Siberia and Alaska. It is also connected to the Arctic Ocean by the Bering Strait, the narrow (53 miles) passage between Alaska and Siberia, once presumed to have been frozen over, allowing the first wandering

Asiatic tribes to find their way to the Americas. Named for Danish explorer Vitus Bering (1681–1741), who first sailed these waters in 1725, the Bering Sea is icebound from November to May.

It should not be confused with the much smaller Barents Sea, a shallow part of the Arctic Ocean to the north of Russia named for a Dutch navigator who sailed there in the sixteenth century. Warmed by currents from the south, the Barents remains ice-free in winter, historically giving it enormous strategic value to the Russians.

Gulf of Mexico Although not called a sea, this vast basin, bounded by Mexico and the American southern states from Texas to Florida, is more than 600,000 square miles (c. 1,560,000 square km) in area, larger than many seas. The rich oil and gas reserves beneath its surface have produced great wealth, but the pollutants related to these industries have also threatened the rich fishing grounds of the Gulf, important both as an industry and for the tourist economy.

Sea of Okhotsk Another large, icy sea off the Siberian coast and north of Japan, this arm of the Pacific is separated from the Bering by the Kamchatka Peninsula, which juts out of Siberia.

East China Sea Another Pacific arm lying between mainland China and the southern parts of Japan. Vast oil deposits were discovered in the area in 1980.

Hudson Bay Although called a bay, it is a large inland sea in northern Canada, connected to both the Atlantic and the Arctic Oceans. The bay is named for the Dutch explorer Henry Hudson, who reached it in 1610 while searching for the Northwest Passage and died there when his men mutinied and set him adrift in a small boat. Hudson Bay freezes over in the winter.

Sea of Japan Lying between the Japanese islands and the Korean Peninsula, this arm of the Pacific contains a warm flowing current that keeps the Russian port of Vladivostok, to the north, free of ice in the winter, the only Russian port on the North Pacific open year-round.

North Sea An arm of the Atlantic lying between Great Britain and Scandinavia, its waters have always been rich fishing grounds. The discovery of gas and oil have also made the area a leading energy producer. With an area of 222,000 square miles (575,000 square km), it is also known as the German Ocean.

Baltic Sea (*Ostsee* in German) An arm of the Atlantic, the Baltic is bounded by the Scandinavian countries and north-central Europe. Shallow and with low salinity, it freezes over during the winter months. The Baltic was thrust into prominence during the breakup of the Soviet Union because it gives its name to the three Baltic states of Lithuania, Estonia, and Latvia, which were part of the Soviet Union until they declared their independence in 1991, largely helping to bring about the demise of the USSR.

NAMES: Yellow Sea, Red Sea, Black Sea: Are They Truly Yellow, Red, and Black?

This is the nautical rainbow coalition. But does each of these seas get its name for very specific color reasons? Well, two out of three isn't bad.

The **Yellow Sea**, or **Hwang Hai**, lies between mainland China and Korea and gets its characteristic color from the rich yellow silt, called loess, deposited by the Yellow (Hwang Ho) and other rivers.

The narrow body of water separating northeastern Africa from the Arabian Peninsula, the **Red Sea** gets its name from the masses of reddish seaweed found in its waters. The Red Sea was created when the Great Rift running the length of Africa opened up and waters from the Indian Ocean rushed in. As the Great Rift continues its slow geologic widening, the Red Sea will become larger. But don't wait around for it to happen.

Before Europeans figured out how to sail around Africa to get to the East, the Red Sea was a major trade connection between the Eastern Mediterranean and the Orient. Its importance was restored when the Suez Canal was opened in 1869, giving it direct access to the Mediterranean. Ships using the Suez–Red Sea route avoided thousands of miles of difficult sailing around Africa. However, the significance of the Red Sea has once again been diminished as most modern super-tankers are too large to navigate the Suez Canal.

The **Black Sea**, much-disputed through history, is a tideless inland sea between Europe and Asia, once called the *Pontos Axeinos* or "inhospitable sea" by the Greeks and *Marea Neagra* by the Romanians. In Greek myth, it was the sea Jason sailed across in his search for the legendary Golden Fleece. Together with the Caspian and Aral Seas, it was once part of a much larger inland sea millions of years ago.

Although its waters are quite dark, its name is believed to be derived from its stormy character rather than any particular color scheme.

The Black Sea is fed by some of Eastern Europe's major rivers, including the Dniester, the Dnieper, and the Danube. The large influx of fresh water creates two levels within the lake. Below a certain depth, little life exists.

The composer Johann Strauss, who wrote the famous "Blue Danube Waltz," would be hard-pressed to recognize his inspiration today. The fabled Danube River originates in southern Germany, then drifts lazily, touching eight countries along the way, before it empties into the Black Sea. But as it moves through Central Europe's industrial heartland, the Danube is continually fed by befouled tributaries, and in turn is a major polluter of the Black Sea.

Geographic Voices   Mark Twain's account of meeting Tsar Alexander II at Yalta on the shores of the Black Sea, from *The Innocents Abroad* (1869)

A strange new sensation is a rare thing in this humdrum life, and I had it here. It seemed strange—stranger than I can tell—to think that the central figure in the cluster of men and women, chatting here under the trees like the most ordinary individual in the land, was a man who could open his lips and ships would fly through the waves, couriers would hurry from village to village, a hundred telegraphs would flash the word to the four corners of the Empire that stretches over a seventh part of the habitable globe, and a countless multitude of men would spring to do his bidding. I had a sort of vague desire to examine his hands and see if they were of flesh and blood, like other men's. Here was a man who could do this wonderful thing, and yet if I chose I could knock him down. If I could have stolen his coat, I would have done it. When I meet a man like that, I want something to remember him by.

As a general thing, we have been shown through palaces by some plush-legged filagreed flunkey or other, who charged a franc for it; but after talking with the company half-an-hour, the Emperor of Russia and his family conducted us through all their mansion themselves. They made no charge. . . .

We spent half-an-hour idling through the palace, admiring the cosy apartments and the rich but eminently home-like appointments of the place, and then the Imperial family bade our party a kind good-bye, and proceeded to count the spoons.

Who Killed the Dead Sea?

First of all, it isn't a sea at all but a lake. The Dead Sea, forming part of the border between Israel and Jordan, is a landlocked salt lake with no outlet. With the Jordan River as its source, the Dead Sea is located 1,289 feet below the level of the nearby Mediterranean Sea, making it the lowest exposed point on the Earth's surface. In biblical times it was known as the Salt Sea because its salt content makes it the saltiest "sea" on Earth. The high salt content is a result of rapid evaporation of the water due to the area's extremely high temperatures.

This extremely high saline level makes it difficult to sustain any life forms, which is why it came to be called the Dead Sea. In the Middle Ages, visitors believed that the air above the Dead Sea was poisonous, because no birds flew over its waters. But there are no birds there because there is nothing for them to eat; there are no plants, and any fish carried in from the Jordan River are killed immediately by the water's high salt content.

The Dead Sea has long attracted tourists for its religious significance and the healthful qualities of its buoyant waters. Today its greatest fame comes from the scrolls associated with the area. First discovered in a cave near Jericho by a Bedouin shepherd boy in 1947, the Dead Sea Scrolls are parchment versions of some of the books of the Bible, many dating from before the time of Christ. Written in ancient Hebrew, and relating contemporary events, the scrolls have offered enormous insight into the authenticity and historical context of the Bible. Selected portions of the scrolls, many of which were in tiny fragments that have been painstakingly reassembled, have been released over the years. But most of the information has remained under the exclusive control of a small group of scholars authorized to study their contents.

There are other reminders of the great religious significance of the area. The Arab name for the Dead Sea is Bahr Lut, or the Sea of Lot, and near the lake's southwest corner stands a low salt mountain that is supposedly the biblical pillar of salt that Lot's wife was turned into after the destruction of Sodom and Gomorrah. Those notorious twin cities of sin were believed to be submerged in the Dead Sea following a volcanic eruption. And on one shore of the Dead Sea is the fabled fortress of Masada, where the Jews made their determined, suicidal stand against the Romans in AD 472.

Although it is a relatively short and small river, the Jordan, source of the Dead Sea, is rich in association for Muslims, Jews, and Christians. Originating near Mount Hermon on the Lebanese-Syrian border, it passes through Lake Huleh (the Waters of Merom in the Bible) and then into Lake Tiberias, a lake in northern Israel also known as the Sea of Galilee, the area most closely associated with the ministry of Jesus.

Where Is the World's Largest Lake?

Like the Dead Sea, the Caspian Sea, situated between Russia and Iran, is a lake. Although lakes are generally associated with fresh water, in strict geographical terms, a lake is any large inland body of water. Completely landlocked, the Caspian Sea is the largest *lake* in the world.

The Caspian, which is about ninety-two feet below sea level, is also the lowest point in Europe.

The nearby Aral Sea is also a lake. With the Caspian and the Black Sea, the Aral once formed an immense, prehistoric inland sea. The Aral was at one time the fourth-largest sea in the world, nearly the size of Ireland. But in recent years, so much of its waters have been diverted to irrigate rice and cotton fields that there has been a rapid drop in the Aral's water level. And as the Aral Sea shrinks, some experts suggest that the climate in central Asia is becoming hotter.

The World's Largest Lakes

The Caspian Sea is on the border of Iran and the former Soviet Union (Russia, Azerbaijan, Kazakhstan, and Turkmenistan all touch the Caspian). Probably named for an ancient tribe called the Caspii, the "sea" is most famous for producing Beluga caviar.

Lake Superior, bordered by Ontario (Canada), Michigan, Wisconsin, and Minnesota, is the world's largest freshwater lake and one of the five Great Lakes.

Lake Victoria (Victoria Nyanza), in the Great Rift Valley of mountainous East Africa, is the chief source of the Nile River and is bordered by the countries of Uganda, Tanzania, and Kenya.

Aral Sea (Aral'skoye More), located in central Asia, east of the Caspian, is surrounded by the former Soviet republics of Kazakhstan and Uzbekistan.

Lake Huron, bordered by Ontario and Michigan, is the second-largest of the Great Lakes.

Lake Michigan is the third-largest of the Great Lakes and the only one completely within the United States. It is bounded by the states of Michigan, Illinois, Wisconsin, and Indiana. (If for some reason you want to remember the names of all of the Great Lakes in size order, just think of SHMEO—Superior, Huron, Michigan, Erie, Ontario).

Lake Tanganyika is a deep lake in east-central Africa, bordered predominantly by the countries of Tanzania and Zaire. The first Europeans to reach it were the explorers Sir Richard Burton and John Speke in 1858 as they sought the source of the Nile. Stanley and Livingstone had their famous meeting near its shores in 1871.

Lake Baikal (Ozero Baykal), located in Siberia, is the world's deepest lake, containing more water than all five of the Great Lakes of North America put together. Tradition holds that Genghis Khan, the famed Mongol emperor, was born near the shores of Baikal.

Great Bear Lake, on the Arctic Circle in Canada's Northwest Territories, is frozen for all but four months of the year.

Lake Malawi (Nyasa Lake), situated in the southern section of Africa's Great Rift Valley in the countries of Malawi and Mozambique, is also known as Calendar Lake because it is 365 miles long and 52 miles across at its widest point.

Great Slave Lake, like Great Bear, is in Canada's Northwest Territories and the capital of the province, Yellowknife, is located on the lake's north shore.

Where Does All the Water Go at Low Tide?

This is one of those childish "Why is the sky blue?" sort of questions to which the answer is not as simple as it seems. The obvious answer isn't necessarily the correct one. Many people probably think that if it is high tide on one side of the ocean—say, the western side of the Atlantic—then it must be low tide on the eastern side of the Atlantic. It isn't that simple. The sea doesn't just slosh back and forth between the two sides of the ocean like water in a barrel that's being rocked back and forth.

Tides are the regular rise and fall of coastal water levels, caused by gravity—stimulated by the attraction of the moon and, to a lesser degree, the sun. The moon, 250,000 miles from the earth, is the major influence on the regular ebb and flow of the earth's tides. Even though the sun is vastly larger than the moon (the sun has *27 million times* the mass of the Moon), its much greater distance from the earth weakens its impact on the earth's tides.

In simple terms, the oceans on the side of the earth facing the moon are "pulled" toward the moon, causing a bulge or high tide. At the same time, the oceans on the opposite side of the earth—facing away from the moon—also bulge in the exactly opposite direction, the result of centrifugal force. These two bulges produce high tides on opposite sides of the earth as water is drawn away. At the same time, there are compensating low tides halfway between the two bulges. As the moon orbits the earth (in the same direction the earth is spinning), these bulges literally "travel" around the earth. One way to picture this is as a rubber band. As you pull two ends of a rubber band, the ends stretch away from each other—that would be high tide. In between, the rubber is stretched thin, that's low tide.

Since it takes the moon a little more than a day to orbit the earth, there are two cycles of tides in roughly every twenty-five hours. From the low point in the tidal cycle, coastal waters rise gradually in the *flood tide*, lasting a little more than six hours. Maximum water level, or *high tide*, is reached and the water begins to drop or move away from the coast in the *ebb tide*. After about six hours, the minimum water level, or *low tide*, is reached and the process begins all over again. There are also regular fluctuations in these cycles. The highest tides, called *spring tides* (even though they have nothing to do with the season), occur twice a month when the sun and moon are in a straight line with the earth. The smallest tides, called *neap tides*, occur when the moon is at a right angle to the sun.

The difference between low and high tide—called the *tidal range*—is different in locations all over the world. In the open ocean, tidal range is insignificant, often no more than three feet. But in coastal shallows, the range is much greater. And in places like bays or channels, where the incoming ocean water is funneled into a narrow inlet, the tidal range is greatly exaggerated. The greatest tidal range occurs in

Canada's Bay of Fundy, which is located between the provinces of New Brunswick and Nova Scotia. There, the tides of the Atlantic Ocean are funneled into a narrow channel to produce a difference between low and high tide of as much as fifty feet (15 meters) twice a day. This dramatic tidal action is now viewed as a potential source of clean, inexhaustible energy. A power station that uses the ebb and flow of tidal water to drive turbines which generate electricity has been in operation for more than thirty years at the mouth of the Rance River in the Gulf of Saint-Malo in France.

Another phenomenon of tides is the *tidal bore*, which, unlike a colossal bore, goes away. On the other hand, like many bores, it comes back. Regularly. A tidal bore is a high wave that travels up an *estuary* (an inlet of the sea where the mouth of a freshwater river meets the incoming salty tide). The most remarkable tidal bore is on China's Qiantang River, which flows into the Bay of Hangzhou. During spring tides, the bore is over twenty-four feet high and travels at almost fifteen miles per hour. The rushing noise of the waters as they move upstream—a sort of reverse waterfall—can be heard fifteen miles away.

What Do Tides Have to Do with Tidal Waves?

In a word, nothing. Until fairly recently, the phrase "tidal wave" was widely used, or, more precisely, misused. "Tidal wave" is a misnomer, an obsolete term inaccurately used to describe either a *tsunami* (Japanese for "overflowing wave"), a fast-moving sea wave caused by an underwater earthquake or volcanic eruption, or a *storm surge*, an abnormal rise in the sea level caused by high winds like those in a tropical hurricane. Whatever you call them, they can be terribly destructive to low-lying coastal areas when they hit. When a cyclone blew out of the Bay of Bengal and hit Bangladesh in 1991, the storm surge and subsequent flooding killed hundreds of thousands of people.

But in two extraordinary recent occurrences, the world witnessed the incredible power of a tsunami, captured live and broadcast around the world. The first was in Indonesia, where an earthquake off the coast of Sumatra on December 26, 2004, created a tsunami wave as tall as 30 meters (98 feet). The most widely recorded tsunami, it resulted in some 230,000 deaths.

The second instance came in March 2011 and struck the North Pacific coast of Japan. Spawned by a magnitiude 9 earthquake, it sent 9- to 10-meter (30 feet) waves sweeping over the east coast of Japan, killing more than 24,000 people and knocking out a nuclear power plant that began leaking radioactive steam.

According to the *Australian Geographic*, the other most deadly tsunamis were:

Ise Bay, Japan (1586)
Enshunada Sea, Japan (1498)
Nankaido, Japan (1707)
Lisbon, Portugal (1755)
Ryuku Islands, Japan (1771)
Northern Chile (1868)
Krakatoa, Indonesia (1883)
Sanriku, Japan (1896)

Less frequent and generally less catastrophic are tsunamis that often accompany a major earthquake or volcanic eruption.

Put simply, a volcano is an opening or vent in the earth's crust, either under the sea or on land, through which rock fragments, gases, ashes, and lava, or molten rock, erupt and are ejected from the earth's interior. There are basically three types of volcanoes; extinct, dormant, and active. The latter two can be incredibly dramatic and destructive while acting as one of the agents of the earth's continual process of destruction and re-creation.

What Is the Ring of Fire?

There are about six hundred active volcanoes on the face of the earth. About half of them range along a belt of active volcanoes running from the southern tip of South America north to Alaska, then west to Asia, and south through Japan, the Philippines, Indonesia, and New Zealand. This is the "Ring of Fire," marking the boundary where the plates that cradle the Pacific Ocean meet those plates that hold the continents surrounding the ocean.

Major Volcanic Eruptions in History

c. 1480 BC—Thera (or Santorini) Located near Crete, in the Mediterranean, this eruption was one of the earliest of recorded volcanoes. Thera was an island outpost of the Minoan civilization (see Chapter 1, "Imaginary Places: Was There an Atlantis?," page 14). The eruption collapsed the island of Thera and sent out a tsunami that probably destroyed much of Crete's economy, leading to the demise of the great Minoan civilization.

AD 79, August 24—Italy One of history's most notorious disasters, the eruption of Mount Vesuvius buried the Roman cities of Pompeii and Herculaneum, killing more than 16,000 residents, who had no chance to escape and died of suffocation. For eight days the black cloud spread over Italy, blocking out the sun as hot rocks and ash fell to the ground in a thick blanket. Located near Naples, Mount Vesuvius is mainland Europe's only active volcano, and it subsequently erupted in **1631**, killing 18,000, **1906, 1929**, and most recently in **1944**, during World War II.

1169—Sicily The island volcano Mount Etna, the highest European volcano, which had erupted many times before, killed about 15,000 people. Another eruption of Mount Etna again killed about 15,000 people in **1669**. Later major eruptions occurred in **1853** and **1928**.

1815, April–July—Indonesia Mount Tambora, on the island of Sumbawa, erupted in one of the largest volcanic blasts ever recorded, darkening the sky at noon with ash from numerous huge explosions. The energy of the blast was estimated to be the equivalent of all the nuclear stockpiles of the 1980s. An estimated 50,000 people living on nearby islands were killed, and as many as 90,000 overall died from the volcano's wider effects. The ash cloud produced was responsible for the havoc in normal weather conditions in the following year, known as the "year without summer," creating severe crop shortfalls in New England and Europe.

1883, August 26–28—Dutch East Indies In May, a dormant volcano on the island of Krakatoa began the most violent and destructive eruption in modern history. The climax came on August 27 with an explosion estimated by modern researchers to have been equivalent to 3,000 Hiroshima-sized atomic-bomb blasts, creating the loudest noise known to man. Violent explosions destroyed three quarters of the island of Krakatoa, sending most of the island up into the air as dust and ash. Most of the estimated 36,000 people who were killed by the volcano were drowned by the enormous tsunamis it produced, some of which reached 100 feet high and raised water levels as far away as England. Dust, ashes, and smoke rose to a height of about 50 miles, circling the Earth and creating unusual red sunsets for years to come. The cloud also blocked the sunlight, causing a worldwide drop in temperatures. The ashes of that explosion helped create a new island volcano, Anak Krakatoa, which first erupted in 1927.

1902, April–May—Martinique Located near the city of Pierre on this Caribbean island, Mount Pelée had last erupted in 1856. When it showed signs of erupting again, island officials were unconcerned. But the volcanic explosion set fire to much of the island, killing almost the entire population of 28,000 in Pierre, many of whom were suffocated by a deadly gas emitted by the volcano.

1912, June—Alaska The Valley of Ten Thousand Smokes on the Alaskan peninsula is named for its many *fumaroles*, or volcanic steam vents. It was created by one of the most explosive volcanoes of the twentieth century when Mount Katmai erupted. Despite its violence, due to its remote location the volcano caused no deaths.

1943, February—Mexico An earthquake originating in a corn-field 200 miles west of Mexico City marked the birth of a vol-cano known as Parícutin. The volcano's cone grew in amazing fashion, attracting researchers and curious onlookers from all over the world who watched as a new volcanic mountain literally grew overnight. After one day, the cone had reached 120 feet in height; after a week, it had grown to 400 feet (125 meters). The flowing lava eventually buried the nearby village that gave the volcano its name, leaving only church spires showing. Amaz-ingly, there were no deaths. By 1950 when the activity stopped, the cone had grown to 7,450 feet (2,270 meters).

1963, Iceland An underwater volcano off the coast of Iceland—itself a volcanic island where geothermal energy provides most of the heat for its population—erupted and created the small island of Surtsey, now a nature reserve.

1980, May—Washington Located in the southwest part of the state of Washington, Mount St. Helens became the first vol-cano to erupt in the continental United States in more than sixty years. Equivalent to a 400-megaton hydrogen bomb, the explo-sion ripped off the top of the mountain and sent it up in a cloud that blew northeastward and dropped ash 600 miles away. A cubic mile of mountain was literally blown away, and floods and boiling mud stripped millions of trees bare for a radius of several miles. Although it was a huge explosion, the area was sparsely populated and the death toll was a relatively low 70. But tens of millions of fish, birds, and animals were also destroyed.

1982, March—Mexico El Chichon, a dormant volcano in Chi-apas, Mexico, suddenly came to life in a series of explosive erup-tions.

1985, November 13–16—Colombia In one of the most deadly eruptions of all time, Nevado del Ruiz caused mud slides that buried most of the town of Armero and devastated the Chinchina River valley, leaving 25,000 dead.

1991, June 10—Philippines Quiet since 1380, Mount Pinatubo on the island of Luzon erupted, covering thousands of miles in ash, dust, and mud slides. There were political ramifications involved, as the United States was forced to evacuate and close Clark Air Force Base.

2010 April—Iceland The ash from the volcano Eyjafjallajokull (pronounced EYE-a-fyat-la-jo-kut-l) shut down European air travel for days, although there were no casualties.

What's the Difference Between an Island, an Islet, and an Isle?

Some people like to pretend that size doesn't make a difference. Well, it does when it comes to islands. This minor piece of geographic confusion is fairly simple to resolve. *Islands* are bodies of land completely surrounded by water. *Islets* and *isles* are basically small islands with no real objective criteria to set them apart.

But wait a minute. If the ocean really surrounds all the land, doesn't that make all the land in question an island? Again, geography isn't an exact science. So it's safe to call any land smaller than a continent, that is surrounded by water, an island. That still leaves Australia with an identity crisis, but Australia can go both ways. So now we have seen lakes called seas, seas called gulfs, and islands called continents. No wonder a lot of people get confused about geography.

Islands come in two main varieties. There are *continental islands*, which were formerly pieces of a continent that have somehow been separated from the mainland. Great Britain is an example, as are the isle of Manhattan and nearby Long Island. (Four of New York City's five boroughs are on islands unconnected to the mainland of the United States; the Bronx is the exception. This confirms the widely held popular notion that New York City isn't really part of America.)

Then there are *oceanic islands*, or islands formed in the ocean independent of the mainland. Many of the Pacific islands, such as the Hawaiian chain, are volcanic, thrown up from the ocean floor. And the Pacific's Ring of Fire is an example of how a seemingly destructive natural force like a volcano is actually part of the earth's creative cycle. The process can take centuries in the case of a constant, steady flow of lava, or may be sudden and dramatic, as with Surtsey, a small island formed off the coast of Iceland in a burst of lava from beneath the Atlantic.

Besides volcanic islands, the second type of oceanic island is a *coral island*, formed out of—you guessed it—coral.

NAMES: *If Greenland Is All Glacier, Why Isn't It Called Iceland?*

The Vikings were terrible at naming things. By all accounts, they landed somewhere in Canada or perhaps northern New England and set up a colony there in around the year 1000. They stayed for two years before pulling up stakes. They called this colony Vinland. It is not a great spot for growing grapes.

But the naming of Greenland, the world's largest island, was less a mistake than an attempt at propaganda. The notorious Eric the Red was a Viking with a long criminal rap sheet—he was banished from his native Norway for manslaughter and outlawed at least twice more for killings in Iceland. Sailing from the Norse colony on Iceland, Eric and his crew came upon Greenland, rich in game, fish, and birds, and similar in its coastline to his native Norway. While 982 was undoubtedly a warmer period and some of Greenland's glaciers had receded, Eric's chosen name for the island was still a stretch. Hoping to attract more settlers, he named the new country Greenland. Some two hundred years later, the Greenland climate grew colder and the glaciers spread. In addition, the island was visited by plague, the European Black Death, and by the end of the fourteenth century, the Viking colony on Greenland was history.

Today, ice covers one tenth of the land surface of the planet. But it covers four fifths of Greenland, sometimes at depths of up to a mile. Only 5 percent of the island is habitable, basically along two coastal strips where the native population, Inuit with some Danish blood, live off the coastal and deep-sea fishing industry. After the Vikings died out, the Danes arrived in the 1720s and Greenland became a Danish colony in 1815. The island was a part of Denmark until 1953. In 1979 it achieved home rule as a self-governing province, but it officially remains a Danish dependency.

The World's Largest Islands

New Guinea Lying in the southwest Pacific just below the equator and to the north of Australia, the world's second-largest island is divided into two parts politically: independent Papua New Guinea, self-governing since 1973, and Irian Jaya, a province of Indonesia. It is not to be confused with Guinea, a country on the west coast of Africa; its African neighbor Guinea-Bissau; or the South American country of Guyana. And there are no guinea pigs there. They originally came from South America.

Borneo The world's third-largest island is the largest of the Malay Archipelago. (*Archipelago* is a word that has changed meanings. Originally, it meant a sea studded with many small islands, and referred specifically to the Aegean. Derived from the Greek *archi*, meaning "most important," and *pelagos*, meaning "sea," the word is used more commonly now to refer to a large group of islands clustered together.)

Madagascar Lying in the Indian Ocean off the coast of southeast Africa, this is an island of extinct volcanoes; high, rugged mountains; and low, fertile coastal plains.

Baffin Situated between Greenland and Canada, the island is named for William Baffin, an English explorer who reached the island and the adjoining bay in 1616.

Sumatra Lying in the Indian Ocean on the equator, Sumatra is part of Indonesia. It is heavily rain-forested, and rich in oil and other minerals.

Honshu The largest of the four main Japanese islands, it is Japan's industrial and agricultural center and contains the country's six major cities as well as Mount Fuji, the volcanic peak that is Japan's highest mountain (12,388 feet; 3,776 meters).

Great Britain Situated off the coast of Europe, this is the main island in the United Kingdom, on which England, Scotland, and Wales are located.

Ellesmere A barren, mountainous Canadian island in the Arctic Ocean off the coast of Greenland, it is notable for Cape Columbia, the northernmost point in North America.

Victoria Another Canadian island located in the Arctic Ocean.

Bikini: Which Came First, the Swimsuit or the Atoll?

The second type of oceanic island is the coral island. The coral polyp is a tiny sea creature that lives in a shell in fairly shallow waters that are warm and clear. When the polyp dies, the softer parts of the body are washed away, but the skeleton is left behind. New polyps grow on the shells of dead ones, eventually forming a great mass of coral. An *atoll* is a coral reef that forms an almost complete circle around a lagoon. The circular coral reefs of most atolls reach deep down into water where no coral can grow. They may have been reefs in shallow water surrounding a volcano. As the island sinks, or the sea level changes, the coral continues to grow. The original volcanic island disappears far below the lagoon, and the reef forms an atoll.

Perhaps the world's most famous atoll is the Bikini Atoll, located in the Marshall Islands in the central Pacific Ocean. From 1946 to 1958, it was the site of the forth and fifth atomic bomb detonations and twenty-two subsequent United States nuclear bomb tests. The Bikini islanders, who had been evacuated in February 1946, began to return in 1972, but the island was again declared uninhabitable because of the high levels of radiation still evident in 1978. The islanders were again evacuated after having been subjected to the largest dose of plutonium ever monitored in a population.

A year after the first test, in the summer of 1947, a different sort of explosion rocked the world. On the French Riviera, women began to wear a skimpy two-piece swimsuit that some unidentified genius decided to call the bikini, after the island. Why the name was chosen is a geographical and fashion mystery. Maybe the effect of the bikini on watchers was thought to be as explosive as an atomic bomb. Another suggestion is that the woman wearing a bikini was left almost as bare as the islands after the bomb blasts. Whichever reason it is, we can only be grateful that the tests weren't done on some other island. Somehow the notion of going down to the beach to check out the latest styles in "guadalcanals" or "eniwetoks" just doesn't have the right ring.

NAMES: Are There Canaries in the Canary Islands?

As with "bikini," this is another chicken-or-egg type of question. The Canary Islands are a group of volcanic islands about sixty-five miles off the coast of northwest Africa. Still controlled by Spain, the principal islands in the group are Tenerife, Palma, Gomera, Hierro, Grand Canary, Fuerteventura, and Lanzarote. The Greeks were the first Europeans to reach them, but it was the Roman historian Pliny (the Elder) who later named them *Insulae Canariae*, or "Islands of the Dogs," because of the many large dogs that lived there. The name *canary* was passed on to the wild finches native to the islands, which must make them "dog birds," the feathered world's answer to bird dogs.

What's the Difference Between a Peninsula and a Cape?

The word *peninsula* comes from the Latin *paene* for "almost" and *insulae* for "island." A peninsula is literally "almost an island," or a piece of land that is surrounded on most sides by water. One way to visualize a peninsula is as a finger of land reaching out to test the water. A cape (derived from the Latin word *caput* for "head") is a head or pointed piece of land that also sticks out into the water. (Small capes can be called spits or points.) One misnamed cape is Cape Cod in Massachusetts, which is a true peninsula. But British sea captain Bartholomew Gosnold, sailing the waters in 1602, christened it Cape Cod.

Other examples of true peninsulas in North America are the state of Florida, Baja California—the narrow strip of land separated from the rest of Mexico by the Gulf of California—and the Alaska Peninsula, which stretches out into the Bering Sea.

Some geographers call all of Europe a peninsula of Asia, which shows that geographic descriptions can be stretched. Within Europe itself, there are several noteworthy peninsulas. Boot-shaped Italy is an example of a pure peninsula, as is the Scandinavian Peninsula, divided into Norway and Sweden. Less typical of true peninsulas are the Iberian Peninsula, occupied by Spain and Portugal, and the Balkan Peninsula, which is divided between the states of Greece, Yugoslavia, Romania, Albania, and Bulgaria. Although both the Iberian and Balkan peninsulas are bodies of land jutting out into the water, neither has the long, narrow shape characteristic of a true peninsula.

Rivers Run, but Can They Drown?

Geographically speaking, the answer is yes. *Rias* are "drowned rivers." How can something made out of water drown?

One of the greatest influences on coastlines is the change in sea levels. When the sea rises, the existing terrain is literally drowned, or covered in water. After the last ice age, in many places around the world, the seas rose significantly, often submerging or "drowning" previous river valleys. Instead of emptying into the ocean, the mouth of the river was submerged beneath the ocean's waters. Chesapeake Bay, on the Atlantic coast of the United States, is one of the world's largest rias. It is the drowned mouth of the Susquehanna River.

Statistically speaking, the rivers of the world are fairly insignificant. Oceans hold more than 97 percent of the world's water, glaciers contain about 2 percent. What little surface water there is on earth—which amounts to the remaining 0.2 percent of the world's total water supply—is mostly locked up in lakes. Rivers and streams contain a small fraction of all the earth's water.

But sometimes statistics are meaningless. It is simple to see that rivers have been central to the development of early civilizations. In particular, the Mesopotamian societies centered on the Tigris-Euphrates, the Egyptian centered on the Nile, and the earliest Chinese centered on the Huang Ho. Initially, these rivers provided fresh water for human and animal consumption, irrigation as agriculture gradually developed, food in the form of fish, and fast, easy transportation.

The World's Longest Rivers

Nile Originating in the mountains of East Africa, the world's longest river begins in two separate streams: the White Nile, with a headstream in Burundi above Victoria Nyanza (Lake Victoria); and the Blue Nile, which rises above Lake Tana in Ethiopia. The two streams meet at Khartoum, Sudan. North of Cairo, Egypt, the Nile empties into the Mediterranean, fanning out into a 115-mile-wide delta.

Amazon With its beginnings high in the Peruvian Andes, the Amazon flows through the world's largest equatorial rain forest, draining 40 percent of South America.

Missouri-Mississippi These two rivers join just above St. Louis, Missouri, to form one of the world's longest rivers. The Missouri, flowing out of the mountains in Montana, is actually the longer of the two. But the Mississippi has historically been more significant as a transportation artery in the development of the American continent.

Yangtze-Kiang (Chang Jiang) China's longest river rises in the Kunlun Mountains of Tibet, running through China's agricultural heartland and providing 40 percent of China's electricity through hydroelectric stations before it reaches the East China Sea at Shanghai.

Ob-Irtyish Situated in Siberia, the Ob and its chief tributary, the Irtyish, are frozen almost half the year.

Huang Ho (Yellow) Rising in the Kunlun Mountains like the Yangtze, the Yellow is named for its loess, or fertile yellow silt. It is also known as "China's Sorrow" because its terrible flooding in the past has been extremely destructive.

Paraná Overshadowed by the Amazon, this Brazilian river is an important commercial artery and the site of the world's largest hydro-electric plant.

Zaire (Congo) Dr. David Livingstone was the first European to explore this African river, perhaps the most famous of the sub-Saharan African rivers. It rises as the Lualaba River in the center of Africa and flows north through Zaire (formerly the Belgian Congo) until it turns in a westward arc and then empties into the Atlantic.

Heilong Jiang (Amur) This river rises in Mongolia and forms the boundary between northeastern China (Mongolia) and eastern Russia.

Milestones in Geography IV
The Nineteenth Century

1807 American inventor Robert Fulton (1765–1815) tests his *North River Steam Boat* on New York City's East River. Later renamed the *Clermont*, it begins regular runs on the Hudson River between New York and Albany. Although not the first steamboat, it is the first practical and economically viable one. In 1819, the paddle steamer *Savannah* became the first steamship to cross the Atlantic. Even though it still relied largely on sail power, its success made steam viable and ushered in the era of steam travel.

1824 American Jim Bridger (1804–81), one of the most celebrated of the famous "mountain men" who traded, trapped, and explored throughout the American West, discovers the Great Salt Lake in modern Utah and thinks its briny water is an arm of the Pacific Ocean. It was also Bridger who opened the Oregon Trail, the principal westward route for settlers coming from the East.

1824 The Erie Canal opens. Begun in 1817, the 365-mile-long waterway connects Albany and the Hudson River to Buffalo and the Great Lakes. Economical boat transportation is now possible between New York City and the Midwest, greatly stimulating the growth and development of both New York and the midwestern cities on the Great Lakes.

1825 The first practical railroad service begins in London as British engineer George Stephenson's *Locomotion Number 1* makes its first trip.

1830 Publication of *Principles of Geology* by Scottish geologist Charles Lyell (1797–1875), a revolutionary approach to concepts of the earth's formation. Lyell laid out the notion that the earth's natural features were the result of a long, continual process, an idea first developed by James Hutton (1726–97) and called uniformitarianism. These ideas upset the prevailing notion that major changes in the geological structure of the earth were the result of abrupt, violent changes (or catastrophes), an idea known as catastrophism.

1831–36 Voyage of HMS *Beagle*, a British surveying ship on which Charles Darwin (1809–82) served as naturalist. During this five-year trip, Darwin begins to formulate his revolutionary theory of natural selection and its role in evolution.

1837 British scientists Charles Wheatstone and W. F. Cooke patent the first electric telegraph, and it is soon used on English railroads and in mapping to determine longitude. In 1838, American scientist—and painter—Samuel F. B. Morse (1791–1872) takes out an American patent for his version of the electric telegraph and the Morse code, both developed with the assistance of physicist Joseph Henry (whom Morse later refused to acknowledge), who was also the inventor of the first practical electric motor (in 1831).

1842–48 John Charles Frémont (1813–90) makes expeditions into the American West with the legendary mountain man Kit Carson (1809–68) as guide. An American soldier, adventurer, and explorer known as "the Pathfinder," Frémont established reliable overland routes to the West and was the high priest of American expansionism. He openly provoked the war with Mexico and served in it. Although he was court-martialed for treason, he was pardoned by President Polk.

1848 The *Map of Oregon and Upper California*, by Frémont and cartographer Charles Preuss, provides a detailed and scientifically accurate map of the American West, spurring a great rush of settlers into the western lands.

1848–49 The California Gold Rush brings tens of thousands of new settlers to the West, both overland using the Frémont-Preuss maps and by sea. After California was taken from Mexico and gained admission to the Union in 1850, Frémont became a senator from the state, then ran as the first Republican presidential candidate (losing to Democrat James Buchanan). He fought for the Union in the Civil War, but was fired by Lincoln. He briefly ran against Lincoln in 1864. A great popular hero, Frémont ended up bankrupt after the failure of a fraudulent railroad scheme.

1848 The Illinois-Michigan Canal opens, linking the Great Lakes to the Mississippi River.

1853 British explorer and writer Sir Richard Francis Burton (1821–90) makes his famous pilgrimage to the Islamic holy city of Mecca disguised as a Muslim. Had his identity been revealed, he would have been killed. His exploits make him one of the the most famous men in England just as his subsequent translations of the *Kama Sutra* and *Arabian Nights* make him one of the most notorious characters of the day.

Geographic Voices   From *Personal Narrative of a Pilgrimage to El-Madinah and Meccah* by Sir Richard Burton

The oval pavement round the Ka'abah was crowded with men, women and children, mostly divided into parties . . . some walking staidly, and others running, whilst many stood in groups to prayer. What a scene of contrasts! Here stalked the Badawi woman, in her long black robe like a nun's serge, and poppy-colored face-veil, pierced to show two fiercely flashing orbs. There an Indian woman, with her semi-Tartar features, nakedly hideous, and her thin legs, encased in wrinkled tights. . . . Every now and then a corpse, borne upon its wooden shell, circuited the shrine by means of four bearers, whom other Moslems, as is the custom, occasionally relieved. A few fair-skinned Turks lounged about, looking cold and repulsive, as their wont is. In one place a fast Calcutta *Khitmugar* stood, with turban awry and arms akimbo, contemplating the view jauntily, as those "gentlemen's gentlemen" will do. In another, some poor wretch, with arms thrown on high, so that every part of his person might touch the Ka'abah, was clinging to the curtain and sobbing as though his heart would break.

1854 After one failed attempt, American commodore Matthew C. Perry (1794–1858) sails into Tokyo Bay and completes a trade treaty that opens Japan to Westerners, ending two hundred fifty years of Japanese isolation.

1855 Scottish missionary and explorer David Livingstone (1813–73), who went to Africa in 1841 as a missionary, reaches the waterfall he names Victoria Falls, located on the Zambezi River in Central Africa on the Zambia-Zimbabwe frontier. A sworn enemy of slavery, Livingstone became the greatest explorer of Africa. He discovered the Zambezi River in 1851 and explored its course during three separate expeditions. Upon his death in 1873, two loyal African followers carried his body fifteen hundred miles so it could be returned for burial in England.

1856 Peak XV in the Himalayas is declared the highest mountain in the world, and is later named Mount Everest.

1858 British explorer Sir Richard Francis Burton (see above, 1853), now internationally renowned, attempts to solve one of geography's oldest mysteries, the source of the Nile. Traveling with fellow explorer John Speke (1827–64), Burton reaches Lake Tanganyika, but it proves to be too low to be the Nile's source. The men are told of another lake higher in the mountains, and Burton, too ill to travel, sends Speke on alone. Speke discovers Lake Victoria and, without adequate exploration or scientific measurements, assumes it is the source of the Nile. The pair become rivals and Speke returns in 1862 to find Ripon Falls, but still fails to verify the lake as the river's source. On the day the two men are supposed to debate the question in London, Speke kills himself.

1869 The "golden spike" is hammered in at Promontory Point, Utah, completing the first transcontinental railway link in America between the Atlantic and the Pacific. (In Canada, a similar link was made in 1885.)

1869 The Suez Canal opens, a 101-mile waterway connecting the eastern Mediterranean to the Red Sea. Completed ten years after work began under the direction of French engineer Ferdinand de Lesseps (1805–94), the canal cuts more than four thousand miles off the sea route from Great Britain to its colony in India. An attempt by de Lesseps to build a canal across Panama ended in financial scandal and de Lesseps, once an international hero, was ruined.

1869 American geologist John Wesley Powell leads the first expedition down the Colorado River through the Grand Canyon. A second expedition in 1871, financed by Congress and better equipped than the first, carefully maps and studies the area.

1870 Mont Cenis Tunnel, the first major railway tunnel, is completed in the Alps.

Geographic Voices   Henry M. Stanley at Lake Tanganyika, 1871

> "Dr. Livingstone, I presume?"
>
> "Yes," said he, with a kind smile, lifting his cap slightly.
>
> I replace my hat on my head, and he puts on his cap, and we both grasp hands, and then I say aloud:
>
> "I thank God, Doctor, I have been permitted to see you."
>
> He answered, "I feel thankful that I am here to welcome you."

These were the words reportedly spoken by Welsh-born soldier, journalist, and adventurer Henry M. Stanley upon meeting Dr. David Livingstone near Lake Tanganyika in 1871. Reported as dead by some of his followers, Livingstone had become the object of an international search and it was Stanley, a brash adventurer working for a New York newspaper, who finally found him. Stanley went on to follow in Livingstone's footsteps as one of the greatest of African explorers.

Known to the natives as Bula Matari, or "Breaker of Stones," Stanley eventually made three epic journeys in Africa. First he crossed the continent from east to west, proving that the Congo and the Nile Rivers were not connected and later verifying Speke's claim for Lake Victoria as the source of the Nile. Working with Belgian King Leopold II, Stanley also explored the Congo region in an attempt to set up a colony there. And finally in his most grueling expedition, he set out with a large rescue party to relieve Emin Pasha, governor of Egyptian Sudan. Supposedly isolated and threatened by the Islamic armies of the Mahdi that had routed the British in the famous disaster at Khartoum, Pasha greeted Stanley's decimated party and informed the explorer he didn't want or need to be rescued.

1872–76 The British steam vessel HMS *Challenger* makes a round-the-world voyage, the first oceanographic survey of its kind.

1872 The United States Geological Survey is founded at the urging of John Wesley Powell. A civilian agency charged with geological and geographical exploration, the USGS will also undertake the mapping of the entire United States.

1883 The *Orient Express*, connecting Paris to Constantinople (Istanbul), makes its first run.

1883 The Brooklyn Bridge is opened to traffic.

1887 German inventor Gottlieb Daimler (1834–1900) installs the internal-combustion engine he had devised four years earlier in a four-wheeled vehicle, one of the first automobiles.

4

Elephants in the Alps

Although we are mere sojourners on the surface of the planet, chained to a mere point in space, enduring but for a moment of time, the human mind is not only enabled to number worlds beyond the unassisted ken of mortal eye, but to trace the events of indefinite Ages before the creation of our race.

—Sir Charles Lyell
Principles of Geology

And so had we been present where the forest meets the savanna one morning five million years ago, we would have caught a glimpse of our ancestors. Still in shadow, they stood peering anxiously across the bright panorama. It would have been easy at a distance to mistake them for a family of chimpanzees. Except that as they started forward through the grass, they kept erect. Each of the adults held a pointed stick in one hand. All of history was there that morning— all that we were to become and still might be.

—Marvin Harris
Our Kind

Are We All "Out of Africa"?
Where Was the World's First City?
Did Moses Part the Red Sea?
Imaginary Places: Was There a Troy?

Geography is history. From the geographic factors that determined the course of evolution to the fact that people built their cities near rivers to all the wars that men have fought to get what was on the other side of the hill, geographic factors have shaped the events that have shaped our world.

Until now, this book has looked at what the world looks like without too much interference from people. Now the emphasis shifts. This chapter sets out to condense about five million years of human history into an overview of the geographic factors that have brought our maps to their present state.

Are We All "Out of Africa"?

Library shelves are groaning with rival paleontologists doing an evolutionary version of Abbott and Costello's famous "Who's on First?" routine that might be called "Who Came First?" The crux of the debate is where on the human family tree certain fossils discovered during the past two decades fit. The bickering between rival camps has not been pretty.

But there are many basic points of agreement. Rounding things out a bit, it is safe to say that human precursors were on the scene four to five million years ago. That seems like a long time. But that time period represents about one thousandth of the earth's life span. Pretty humbling stuff.

And none of the rival bone-hunting camps dispute the notion that those precursors came from Africa. To simplify matters, the primate and human fossil record goes something like this. Human precursors called *australopithecines* ("southern apes") seem to have emerged in Africa four to five million years ago. By 2.5 million years ago, these upright walkers were at home in both trees and on the ground and were probably using crude tools made of bone and stone. Perhaps the most famous of them is paleontology's superstar, "Lucy," a remarkably complete skeleton found in 1974 by a team led by Donald Johanson. Dated to more than three million years ago, Lucy was also known as *Australopithecus afarensis*, because the skeleton was found in the Afar region of Ethiopia. In 1984, another team working in Kenya found the jawbone of an *afarensis* that was dated to five million years ago, the oldest known representative of the hominid line.

Why Africa? Geography has everything to do with this. To paraphrase bank robber Willie Sutton, who said he hit banks because that's "where the money is," Africa is where the primates were. And the evolution of the primate behavior and physiology that finally produced humankind was fostered by the geographic factors of climate and topography. During this much cooler time in the earth's history, the Sahara was a fertile region teeming with large-animal life, and much of the African rain forest was replaced by grasslands, where the ability to walk upright was advantageous for seeing across the open savanna. The open-grassland scene of Pleistocene Africa is a stark contrast to the other places where you find primates, such as in the rain-forest settings of South America.

Schoolchildren used to learn that tool-making was one of the attributes that sets humans apart. But scientists working in both the field and in zoos have observed a wide variety of animals who craft and use rudimentary tools, from elephants who scratch their backs on trees to chimps who hunt for ants and termites with pointed sticks. But the ability to craft and transport tools over any distance begins to separate human behavior from other animals. Combine the advantage of being able to carry tools or rudimentary weapons while walking upright and you have the basic recipe for people.

By about 2.5 million years ago, one line of "southern ape" evolved to produce a more advanced creature, a scavenger named *Homo habilis* ("dexterous man" or "handy man") who possessed a greatly improved tool kit—although it was still a long way to Black & Decker.

More advanced still was *Homo erectus* ("upright, or erect, man"), who started out as early as 1.6 million years ago and died out as recently as 300,000 years ago. Unlike the popular image of a bent-over, apelike creature, *Homo erectus* may have stood six feet tall, judging from a recent fossil discovery of a young *erectus* male. Being tall and thin made evolutionary sense in the heat of savanna Africa because such a body cools more efficiently. It is a good body for both running away and chasing things over long periods of time. The *erectus* could cool himself off more easily than either his prey or the animals that might have enjoyed making a dinner of him. For a perfect example of natural selection at work, contrast this body type with that of a native of the colder Arctic regions, where evolution has favored the development of short, fat body types that retain heat.

Homo erectus spread from Africa into Europe and Asia, reaching as far as China, where the 500,000-year-old "Peking man" was found, but whose remains were unfortunately lost during World War II. Anthropology assumes that *erectus* was responsible for originating the use of fire and the ax. Just give "erect man" a raincoat and a helmet and you could call him "fire man."

Somewhere between 200,000 and 400,000 years ago—exactly when is hard to say; the fossil record is sketchy on this point—good old *erectus* evolved into *Homo sapiens* ("wise man"). Now we're close to modern man, but not quite there yet.

Pause to consider the fascinating and largely unsolved mystery of an Ice Age near-cousin to modern humans called Neanderthal, who was on the scene in Europe and the Middle East by about a hundred thousand years ago, most likely a descendant of *erectus*. One of the first Neanderthal skeletons was found in 1856 in the Neander Valley near Düsseldorf, Germany, which is how this mystery guest got his name. Once presumed to have been both stooped and stupid, the Neanderthal has come up in the world, at least in the eyes of anthropology. In the first place, that Neanderthal skeleton upon which so much theory was based wasn't stooped over because he was apelike but because he suffered from crippling arthritis.

In fact, the Neanderthals may have lived in a fairly complex society that employed elaborate rituals, including ceremonial burial of the dead. At a Neanderthal site in Siberia, the remains of a young boy were discovered beneath a collection of antlers seemingly arranged as a protective shelter for the body. A Neanderthal amulet made from a mammoth's tooth found in modern Hungary has been called the earliest known decorative artifact. But other anthropologists, such as Marvin Harris, are more skeptical about the advances of our Neanderthal cousins. In his fascinating book *Our Kind*, Harris raises the possibility that chance and convenience, rather than the deliberate actions of these early humans, were behind these apparent Neanderthal rituals and cultural artifacts. Harris also doubts that the Neanderthal had developed language, an evolutionary adjustment that clearly set us apart from our predecessors. What we do know is that the Neanderthal disappeared from the picture right after modern humans showed up.

The other great controversy arises over exactly where modern man developed. The roots of the modern human, *Homo sapiens sapiens*—or very wise man, though not very humble—have been traced to Africa, where there is evidence of part-modern, part-archaic humans as far back as a hundred twenty-five thousand years. In caves on the seacoast of South Africa, remains of modern humans who scavenged an existence off shellfish and small animals have been dated at least that far back. And the fossil finds for the conclusion of an "African genesis" have recently been buttressed by more controversial research by microbiologists who have traced a trail of human DNA back to Africa. The prevailing view is that modern humans began to spread out from Africa over tens of thousands of years, gradually supplanting more archaic forms, including the Neanderthal. This fossil record eliminates the possibility that we present humans descended from the Neanderthals, since modern human beings were on the scene in several areas long before Neanderthals arrived in those regions.

But not everyone is satisfied with the contention that we are all "out of Africa." The DNA evidence supporting the "African Eve" has been shown to be flawed. And there is another view. Paleontologists studying skulls found in a variety of areas argue that modern humans may very well have developed independently in several places all over the world. This central debate is a seismic fault line splitting paleontology today.

While this essential mystery may eventually be cleared up with new fossil finds or advances in microbiology that allow far more sophisticated analysis of the genetics of human evolution, some answers are easier. It is certain, based on dating of bone fragments from central Australia, that early humans reached the island continent by 60,000 BC. In Europe, the first modern humans, called Cro-Magnon, lived from about forty thousand years ago in a period of time known as the

Upper Paleolithic. Another ice age had come, and they adapted well to it. In successive waves, some of them apparently crossed into North and South America either by way of the ice-age land bridge that once connected Siberia and Alaska or perhaps even by boat. While a date of twelve thousand years was long accepted as the time frame for humans in the Americas, carbon testing of fragments found in caves in Chile has pushed that date back to as much as thirty-three thousand years ago. More radical, but largely undocumented, theories contend that humans were in the Americas even earlier—another mystery waiting for evidence to fill in some large blanks.

But back to the Neanderthal. For a period of time the Neanderthals and the Cro-Magnon people overlapped in Europe and the Middle East. Then the Neanderthal trail goes ice-age cold. Whether the Cro-Magnon and other modern *Homo sapiens* flat out killed off their Neanderthal neighbors, forced them into less hospitable surroundings where they withered out of existence, or intermingled with the Neanderthal until they were absorbed into the Cro-Magnon gene pool, is still a part of the mystery of human evolution. But we do know that the Cro-Magnon and other modern people went on. Somewhere around thirty thousand years ago, in a variety of places around the world, these prehistoric ancestors made a quantum leap in civilization. They invented art. Elaborate cave drawings of animals and astonishingly sophisticated carved figures—both fertility figurines and delicately carved animals—burst onto the scene. And by around ten thousand years ago, these ancestors had begun to live in villages, to keep animals, and to trade and farm.

Geographic Voices   From *Out of Africa* (1937) by Isak Dinesen (Baroness Karen Blixen, 1885–1962)

I had a farm in Africa, at the foot of the Ngong Hills. The Equator runs across these highlands, a hundred miles to the North, and the farm lay at an altitude of over six thousand feet. In the day-time you felt that you had got high up, near to the sun, but the early mornings and evenings were limpid and restful, and the nights were cold.

The geographical position, and the height of the land combined to create a landscape that had not its like in all the world. There was no fat on it and no luxuriance anywhere; it was Africa distilled up through six thousand feet, like the strong and refined essence of a continent. The colours were dry and burnt, like the colours in pottery. The trees had a light delicate foliage, the structure of which was different from that of the trees in Europe; it did not grow in bows or cupolas, but in horizontal layers, and the formation gave to the tall solitary trees likeness to the palms, or a heroic and romantic air like fullrigged ships with their sails clewed up, and to the edge of the wood a strange appearance as if the whole wood were faintly vibrating. Upon the grass of the great plains the crooked bare old thorn-trees were scattered, and the grass was spiced like thyme and bog-myrtle; in some places the scent was so strong, that it smarted in the nostrils. . . . The views were immensely wide. Everything that you saw made for greatness and freedom, and unequaled nobility.

Where Was the World's First City?

With the recent discoveries of skeletal remains in caves in the Middle East by a joint Israeli-French archeological team, the dates of the earliest "modern" humans have been pushed back to approximately 100,000 years ago. Nomadic hunter-gatherers, these folks would take some time to settle down. About 90,000 years, give or take a few centuries.

The first permanent human settlements were made in the Middle East, with the establishment of sedentary bases with houses, storage facilities, and tools that have been found in modern-day Israel. Remains from this very ancient human era were first found in the Wadi en-Natuf area in modern Israel near the shores of the Mediterranean, and the period has been called Natufian. Permanent settlements next grew in what is now Syria, the Euphrates area (modern Iraq), Persia (modern Iran), and Anatolia (modern Turkey), as hunter-gatherers traded in their flint spears for plowshares and learned how to farm about 10,000 years ago. Çatal Hüyük, located in southern Turkey, was a town of some 6,000 people occupied from about 7000 BC, where some of the oldest known pottery, textiles, and plastered walls have been discovered.

While these agricultural settlements represented a great leap toward civilization, the first true cities were the ancient capitals of the early civilizations, such as Ur of the Chaldees in Mesopotamia, which was founded approximately 4000–3500 BC. The capital of the Sumer kingdom, birthplace of writing and the presumed home of the biblical patriarch Abraham, Ur was once a great city featuring one of the ziggurats, or brick pyramid temples, built in the Mesopotamian region starting around 3000 BC. (One of these ancient pyramids is presumably the basis for the biblical Tower of Babel.) An early trading center, the city was later abandoned when the Euphrates changed its course. At that point, around 2000 BC, Babylon became the capital of the Old Babylonian empire and quickly became the greatest city of the period, featuring the famed Hanging Gardens (see page 168, "What Were the Seven Wonders of Antiquity?").

Thebes and Memphis, at various times the capital cities of ancient Egypt, were probably founded around the same time as Ur. And in the Indus Valley (modern-day Pakistan), Harappa and Mohenjo-Daro (Hindu for "mound of the dead") were sophisticated cities of some 40,000 people that date from about 2500 BC and which traded with Ur. By contrast, the first circular earthen mounds dug at the Stonehenge site in Great Britain date to about 2750 BC, with the stone megaliths apparently put in place by 1700 BC. Such great cities as Cairo (founded as El Fustat in AD 642) and Baghdad (founded about AD 762) are relative newcomers.

The title of the oldest, continuously inhabited city belongs to the Syrian capital of Damascus, the oldest capital city in the world. Built in an oasis and dating to about 2000 BC, Damascus was an early commercial center as a halt for the desert trading caravans. It is mentioned in the book of Genesis in connection with Abraham, the patriarch revered by Jews and in the Islamic world as well. Through its long history, Damascus has been controlled by Assyrians, Macedonians, Romans, Arabs, Mongols, Turks, British, and French. It became the capital of independent Syria in 1946.

But more ancient than any of these early settlements is Jericho, a village located fourteen miles from Jerusalem in Jordanian territory now occupied by Israel. Built 825 feet below sea level, Jericho has been shown in recent excavations to have consisted of at least twenty successive layers, each indicating a different period of settlement. The earliest of these has been dated to approximately twelve thousand years ago. The original site of Jericho is a mound near modern Jericho called Tell es-Sultan, built on an oasis on the desert's edge. Mud-brick houses there were surrounded by a stone wall, parts of which have been recently excavated.

Of course, most people recall Jericho for another reason: the familiar biblical tale of Joshua, the successor to Moses as leader of the wandering tribes of Israel. At this point in history—estimated to be around 1100–1200 BC—Jericho was a town belonging to the Canaanites, the people living in the land promised to Moses and the Israelites by God. In the biblical version, Joshua commands his priests to blow their trumpets, the children of Israel issue "a great shout," and the walls of Jericho "fell down flat," allowing the Israelites to enter victorious and butcher everyone in Jericho, "both men and women, young and old, oxen, sheep and asses, with the edge of the sword"—except for a harlot with a heart of gold named Rahab who had sheltered two Israelite spies in her home.

It makes a great story, but more likely an earthquake caused those walls to come tumbling down. At least that is what one researcher has determined. A Stanford University geophysicist who has been studying the ten-thousand-year-long historical record of earthquakes in the Holy Land, Amos Nur, says that quakes have repeatedly destroyed Jericho, most recently in 1927. The town lies on the Jordan Fault, dividing the Arabian plate from the Sinai plate. Nur's theory is supported by archeologists who have learned that Jericho's walls often collapsed in a single direction, as they would in a quake, not in all directions as if an army had destroyed them. The book of Joshua also says the Jordan River stopped flowing, allowing the Israelites to cross. This typically happens as a result of quakes in the Dead Sea Fault Zone, as the Jordan's banks collapse and briefly dam it.

Did Moses Part the Red Sea?

Of course, to treat biblical events with an objective scientific or historical eye is to tread on thin ice. The risk of being called a heretic is still real. Although the consequences are far less serious than might have been the case, say, around 1492, when a fellow named Tomás de Torquemada was throwing the Jews out of Spain, burning people at the stake, and stretching them on the rack for much less heresy than this.

Despite the risks, science presses on. One example came with the report of a team of oceanographers who had found a way to explain the biblical parting of the waters by Moses, described in Exodus and immortalized by Cecil B. DeMille in *The Ten Commandments* with Charlton Heston raising his staff to part the Red Sea. A theory proposed in 1992 by Dr. Doron Nof, an oceanographer, and Dr. Nathan Paldor, an expert in atmospheric sciences, offers the possibility that strong winds blowing along the narrow, shallow Gulf of Suez could account for the phenomenon.

The salient passage is from Exodus and tells how Moses led the children of Israel out of bondage in Egypt, the most important event in Jewish history. According to the biblical version, "the Lord caused the sea to go back by a strong east wind all that night, and made the sea dry land, and the waters were divided. And the children of Israel went into the midst of the sea upon the dry ground; and the waters were a wall unto them on their right hand and on their left."

When the Israelites had safely passed, according to the account, the water closed in on the pursuing chariots of the pharaoh. The Jews kept going, wandering in the wilderness for forty years, during which time Moses received the Ten Commandments.

Part of the problem for historians and scholars interested in this question has been identifying exactly where and when Moses parted these waters. Dating is difficult, but most scholars accept a date somewhere between 1250 and 1350 BC. For a long time, the passage presumably referred to the Red Sea, or more precisely the Gulf of Suez, an arm of the Red Sea that separates Egypt from the Sinai Peninsula. But in the past thirty years, biblical scholars have reinterpreted Hebrew texts of the book of Exodus as saying the Israelites crossed the "Sea of Reeds," a marshy area at the northern end of the Gulf of Suez, not the Red Sea itself. Other Old Testament historians have suggested a spot even farther north at Lake Manzala, on the Mediterranean coast.

But Nof and Paldor point to the Gulf of Suez, which is narrow and fairly shallow, with mountains on either side of the gulf capable of channeling the winds. The two scientists showed that a strong, steady wind blowing for ten to twelve hours could push water from the shoreline and expose an underwater ridge. They refer to the biblical account, which describes a strong wind blowing for the entire night before the Israelis made the crossing, as being entirely consistent with their theory.

Another proposed explanation for the parting of the waters was the huge sea wave caused by the volcanic eruption of Mount Thera. (See Chapter 1, page 14, "Imaginary Places: Was There an Atlantis?") But the date of that catastrophe seems to be too early to have had any connection with Exodus. If you want to hedge your bets, you can always say that God caused the volcano, or earthquake, that knocked down the walls at Jericho or the winds that parted the seas. That's essentially what Dr. Paldor told the *New York Times*. "Believers can find the presence and very existence of God in the very creation of the wind with its particular properties, just as they find it in the establishment of a miracle."

No one can prove you wrong, and you're safe should the day come when you have to meet your Maker!

IMAGINARY PLACES: *Was There a Troy?*

Certainly less controversial than questioning the veracity of the books of Exodus and Joshua are questions about the sources of Greek mythology and literature. Ironically, we know more about the events described in Homer's epic *Iliad* and its companion *Odyssey*, the starting point of Western literature, than we do about the historical fall of Jericho or the actual Exodus, even though these events may have been separated from each other by only a few hundred years. For that matter, we now also know more about Troy than we do about the presumed author of these poems, a blind poet who recited these stories around 800 BC, about four hundred years after the events they describe took place.

Homer's Troy (also known as the Greek and Roman city of Ilium, which is where the name *Iliad* comes from) was a coastal city of ancient Asia Minor. Around 1250 BC, a war was fought there between the Trojans and an alliance of Greeks. According to the Homeric story, the Greeks sought to avenge the abduction of Helen, the wife of Sparta's

King Menelaus, by Paris, the son of Troy's King Priam. Recited for generations until it was finally written down, Homer's epic recounts, in the words of historian Barbara W. Tuchman, "ten years of futile, indecisive, noble, mean, tricky, bitter, jealous and only occasionally heroic battle." It is a universal, human story of the two camps, their respective heroes, and the often petty involvement of the Greek gods of Mount Olympus who took such an active part in the war's outbreak, fighting, and eventual outcome.

Troy falls when the Greeks leave behind the wooden horse in which Greek soldiers have hidden themselves. The horse is brought inside the walls of Troy, the concealed Greeks let in the rest of their army, and Troy is sacked and burned. (Among Troy's survivors is the warrior Aeneas, who went on to found Italy, a story recounted in the *Aeneid*, the Roman epic by Virgil.) The ploy of the wooden horse was the work of Odysseus (called Ulysses by the Romans), whose adventures in wandering for ten years were told in Homer's *Odyssey*.

To generations of Greeks, these poems provided a cultural heritage almost as sacred as the Bible itself. But the city of Troy remained a mythical place until it was rediscovered between 1870 and 1890 by Heinrich Schliemann (1822–1890), the self-trained son of a poor Protestant minister from a village in northern Germany. He was convinced Troy was buried near Hisarlik, located south of the Dardanelles (Çanakkale Bogazi in Turkish), a narrow strait that connects the Aegean Sea to the Sea of Marmara. While Schliemann's digs at Hisarlik certainly uncovered the site of Troy, he mistakenly dug through several layers of Troys, even passing through the one he was looking for. Eventually, eight layers, built one on top of another, were excavated. The earliest site, Troy I, dates from around 3000 BC. The next level, Troy II, contained an imposing fortress, and its inhabitants apparently made wide trade contacts. Its famous treasure of gold, copper, and bronze indicates a wealthy community. Troy VI, dated around 2000–1300 BC, had a citadel surrounded by huge limestone walls and large houses built on terraces. It was apparently destroyed by an earthquake.

Homeric Troy was probably the rebuilt Troy VII. It was shown to have been looted and destroyed by fire.

What Were the Seven Wonders of Antiquity?

Some may have existed only in myth and legend, perhaps real objects embellished by generations of tall tales. Some are still standing. Pieces of others are preserved as relics in museums. These seven places were listed by Greek writers as the greatest structures of the ancient world. All seven would have been as well known in the world before the time of Christ as the Empire State Building, the Eiffel Tower, the Taj Mahal, or the Sistine Chapel are today, landmarks instantly recognizable to educated people.

The Pyramids of Giza The only one of the wonders still standing, the three great pyramids of Egypt, located outside modern Cairo, were built about forty-six hundred years ago. The largest pyramid, built by Khufu (or Cheops) had an estimated original height of 482 feet (it has eroded to approximately 450 feet). The sides of the base average 755 feet (230.12 meters) in length. The Great Pyramid was built of more than two million blocks, each weighing about two and a half tons. Its builders were able to align it as a nearly perfect square with the sides almost precisely facing the four cardinal points of the compass. The second pyramid, only slightly smaller, was built for Pharaoh Chephren. The third and smallest is the Pyramid of Mycerinus. In addition to the Giza pyramids, the nearby Sphinx, cut from solid rock, is another of Giza's extraordinary features. It bears the head of the Pharaoh on top of the body of a lion.

The Hanging Gardens of Babylon The glory that once was Babylon has been reduced to a few ruins near the Euphrates River in Iraq. But four thousand years ago, Babylon was the capital of one of the great empires of the ancient world. With the defeat of the Assyrians around 626 BC, Babylon became the capital under two emperors, Nabopolassar and his son Nebuchadnezzar II (circa 630–562 BC). Nebuchadnezzar's wife, Anuhia, was the daughter of the Persian king, and according to legend, she missed the hills of her Persian homeland. To appease his queen, Nebuchadnezzar II had the Hanging Gardens constructed about 600 BC.

If reports of their height and beauty can be accepted, this series of terraced gardens rising to a height of 328 feet (100 meters)—three-quarters of the height of the Great Pyramid—was an extraordinary architectural and engineering accomplishment. Wells and hydraulic pumps manned by slaves raised the water needed to maintain the gardens' splendor.

Besides having built the gardens, Nebuchadnezzar II is notorious as the Babylonian ruler who partially captured Jerusalem in 597 BC and then destroyed the city, taking thousands of Jews captive to Babylon in the year 586 BC. But in 539 BC, twenty-two years after Nebuchadnezzar's death, the Persian King Cyrus conquered Babylon.

The Zeus at Olympia Nothing survives of this famed statue of the mightiest of the Greek gods (adopted by the Romans as Jupiter). Completed about 435 BC, it was supposedly forty feet tall and carved of ivory and gold, with Zeus seated upon a great cedar throne. The statue was not located on Mount Olympus, the mythic home of the gods in northeast Greece, but in the Great Temple in the Plains of Olympia in southern Greece, where the ancient Olympic games were held every fourth year.

The mad Roman emperor Caligula wanted to carry the statue to Rome and replace Zeus's head with a carving of his own face. That plan was thwarted when the workmen who had come to move the statue were driven away by loud laughter in the temple. Another Roman emperor, Theodosius, did have the statue moved to Constantinople, where it was destroyed by fire in AD 475. The site of the Great Temple still exists, but only scattered blocks of stone remain. Although nothing remains of the *Zeus*, other pieces of original sculpture from the temple are displayed in the Olympia Museum.

The Temple of Artemis (Diana) at Ephesus Completed about 323 BC, this was supposedly the most beautiful of the ancient wonders, built to honor the daughter of Zeus and sister of the sun god Apollo. Ephesus was an ancient Greek city on the Aegean coastline in what is now Turkey. It was also a Roman provincial capital and is the city in which the Apostle Paul lived, nearly causing a riot when he preached against the licentious worship of Artemis, an ancient fertility goddess.

With 127 marble columns, each 60 feet (18 meters) high, the temple was large and architecturally astonishing. According to one ancient writer, the temple "surpasses every structure raised by human hands." Destroyed in AD 262 by Goth invaders, the temple site was later buried

beneath a river whose course had changed. Some remnants of its renowned columns are preserved in the British Museum.

The Tomb of King Mausolus at Halicarnassus Located near the
modern coastal Turkish city of Bodrum, this monument was erected by
Queen Artemisia in memory of her brother, King Mausolus of Caria,
a province of the Persian Empire in Asia Minor, who died in 353 BC.
Artemisia also died before the work was complete. The finished shrine
in honor of Mausolus was a pyramid resting on a square base. Atop the
pyramid was a sculpture of a horse-drawn chariot in which the king
stood.

Some remains of the structure, including a massive statue of Mausolus, are also in the British Museum. And the shrine is the source of
the word *mausoleum*.

The Colossus of Rhodes Located 12 miles (19 km) from the coast
of Turkey, the island of Rhodes is today a part of Greece. In the year 312
BC, a war was fought between Rhodes and the Greek state of Macedon.
After Rhodes survived an attack, a statue in honor of the god Apollo
was commissioned, to be cast from metal taken from captured Mace-
donian weapons. Built by the sculptor Chares, the colossal statue took
twelve years to complete. When finished in 280 BC, the Colossus representing Apollo towered over the harbor at Rhodes, 105 feet (32 meters)
tall, falsely said to be high enough for ships to sail between its legs.

According to legend, Chares thought he had made a mistake in the
figure's proportions and killed himself. Although well known throughout the ancient world, the Colossus was short-lived. About fifty years
after its completion, the statue was destroyed by an earthquake. The
pieces were still there in the 1st century AD, according to the Roman
historian Pliny, who described their great size.

The Pharos Lighthouse Built in 270 BC on the small island of
Pharos off the coast of Egypt, the lighthouse stood outside the port
of Alexandria, the world's cultural capital since its founding in 332 BC
by Alexander the Great. Designed by the Greek architect Sostratus,
the white marble lighthouse stood 440 feet (1,234 meters) high, with a
square base, an octagonal middle section, and a circular top section. A
fire burned continuously in a brazier at the top of the lighthouse to provide a beacon for Mediterranean shipping. Inside the top was a mirror
that was supposedly used as a weapon, capable of focusing the sun's rays

to set fire to hostile ships. For a long time, the word *pharos* was used to describe any lighthouse. The Pharos Lighthouse continued to serve its function for nearly nine hundred years until the Arabs conquered Alexandria and half-dismantled it. A great earthquake in 1375 shattered the remainder, throwing the great hulks of marble into the harbor, where they remained.

Of course, the Greek writers and historians who recorded these wonders could only discuss things in their known world. Elsewhere in the inhabited world before Christ's time there were other great wonders. The most obvious absentee on the list of the wonders of antiquity is the Great Wall of China, the world's longest wall fortification and perhaps the greatest building project ever undertaken. Although much of the present Great Wall dates from AD 1420 when it was enlarged during the Ming dynasty, the wall originally dates to around 221 BC. That was when the powerful emperor Shih Huang Ti unified the empire under the Ch'in dynasty. To protect his northern border from attack by Hsiungu (Huns), Shih Huang Ti ordered the existing town walls of the northern frontier linked to create the Great Wall. Accounts from that period state that three hundred thousand men labored for years and that the fourteen-hundred-mile-long wall was ready by 215 BC.

Also ignored was an entire city that must have ranked as a wonder, the Persian capital of Persepolis. Given the way the Greeks felt about the Persians, it is no surprise that the Greek writers left out a Persian accomplishment. These were the superpowers of their day and there was no love lost between them. Successor to the empires of Babylon, Egypt, and Assyria, the Persian Empire, based in what is now Iran, was larger and more populous than any of its predecessors of ancient times. Begun by Darius I, who reigned from 522 to 486 BC, Persepolis was built from scratch, a new capital city, anticipating such similarly planned capitals as Washington, DC, and Brasilia.

The site for Persepolis was a natural rock terrace backed by a sheer cliff face. Unlike other ancient building projects such as the pyramids, it was built by paid laborers rather than slaves. Clay tablets from the construction period reveal accounts of the wages paid to the laborers who built the city. An awesome piece of architecture, Persepolis contained huge palaces, ceremonial stairways, and imposing sculptures such as the great carved gateway of Xerxes I, the son of Darius I.

The Persian Empire was at its height as the Greeks began their ascendancy, and the two empires fought a series of bitter wars. Darius was defeated at the famed battle of Marathon (490 BC). His son, Xerxes I, continued the war against the Greeks and was checked at Thermopylae (480 BC), but still destroyed Athens. The Athenians defeated his fleet, however, at the great naval battle at Salamis. Had these Persian wars, recounted by (among others) Herodotus, the "Father of History," gone the other way, it might have been a different ancient world. With the Greek victory began the domination of Greek and hence Western civilization, which from this early beginning expressed a sense of moral and cultural superiority over the East. (See "World Battlefields That Shaped History," page 182.)

The Persian Empire, struggling under court intrigues, was finally subdued by Alexander the Great in 330 BC, when he sacked and destroyed Persepolis in revenge for the earlier destruction of Athens by Xerxes.

In 1971, the shah of Iran—later deposed by the Iranian Revolution of 1979—staged an enormous celebration in the ruins of Persepolis to commemorate the 2,500th anniversary of Iranian monarchy.

Was Cleopatra Black?

In Michael Jackson's Egyptian music video *Remember the Time*, the pharaoh is played by Eddie Murphy and his queen, Nefertiti, by the model Iman. While it is safe to say that the Egyptian pharaohs did not look like Yul Brenner, it may be a leap to think they looked like the star of *48 Hours* either. But it's a sure bet that Moses had little in common with Charlton Heston and Jesus didn't look like Max Von Sydow, the Swedish actor who portrayed Christ in *The Greatest Story Ever Told*. From *Birth of a Nation* to *JFK*, Hollywood has always had its own versions of reality and they have little to do with historical veracity. But if Cleopatra didn't resemble Elizabeth Taylor, who did she look like?

This isn't simply an idle question. The issue of what ancient people looked like, who they were, where they came from, and what they accomplished is at the center of an intense debate in American education that is rattling the sherry glasses in the ivory-white towers of academia. At its heart, the question is about Afrocentrism, a very specific branch of the multicultural movement that wants contemporary schoolbooks to reflect the history, achievements, and contributions of overlooked minority groups alongside those of white European males traditionally taught as "Western civilization."

Afrocentrists point to the specific impact that Africans had on the rise and development of early cultures. This gets to the heart of the Cleopatra question. Some scholars have come to believe that Egypt, held up for so long as the first great civilization, was in fact a black African society. By extension, the Egyptian influences on Greece—and they are many, including Greek mythology, the development of Greek mathematics, and astronomical observations—were African influences. Champions of Afrocentrism point an accusing finger at centuries of biased European scholarship and schoolbooks that they claim have deliberately overlooked the African legacy. Proponents of multicultural and Afrocentric schooling in particular also note the existence of powerful cultures existing in Africa while Europe was in the midst of the Dark Ages, societies completely overlooked in traditional discussions of the medieval world.

The seminal and most scholarly serious work on this question comes from Cornell professor Martin Bernal, whose two-volume *Black Athena* opened this debate. In his award-winning books, Bernal uses ancient documents, archeological evidence, and his considerable expertise as a linguist to trace the influence of Egypt on ancient Greece and, from there, the rest of Western civilization.

Historically speaking, Egypt may have originated in the black African societies of the Upper Nile, in what is now Ethiopia. But over thousands of years, as sub-Saharan Africans mingled with Asian and Mediterranean people, the population became thoroughly mixed. So while Michael Jackson's *Remember the Time* might be a terrific dance tune, it's on very shaky historical ground.

As for Cleopatra—there were actually several Cleopatras during the Ptolemaic period in Egypt. The most famous of them was alive from approximately 69–30 BC and ruled as queen of Egypt from 51 BC. A member of the Ptolemaic family, she was descended from the

Greek dynasty installed by Alexander the Great's generals after they conquered Egypt. Given her ancestry, she most certainly was not black.

But she certainly wasn't boring. First she married her brother and was forced into exile. Then she attracted the attention of Julius Caesar, who restored her to power when he conquered Egypt in 48 BC. As his consort, Cleopatra bore Caesar a son and lived with him in Rome. She returned to Egypt and married another brother but had him poisoned. Enter the Roman general Mark Antony, with whom she lived for twelve years—a bad career move for him, as it cost him support in Rome and he ended up committing suicide after losing the Battle of Actium. When Cleopatra failed to win the love of Antony's rival, Augustus, she supposedly committed suicide by holding a poisonous asp to her breast.

Milestones in Geography V
1900–1949

1903 The first successful airplane is launched at Kitty Hawk, North Carolina, by the Wright brothers, Wilbur (1876–1912) and Orville (1871–1948). The longest flight of the day lasts 59 seconds and travels 852 feet—a speed of 30 mph. Five years later, Orville makes the first flight lasting one hour.

1904 The Russo-Japanese War ends with the Treaty of Portsmouth (New Hampshire), negotiated by American president Theodore Roosevelt. Under its terms, Japan is granted major territorial concessions, greatly expanding Japanese power in the area.

1905 American scientist Daniel Barringer proposes that a large crater in Arizona was caused by a meteor's impact, not a volcano. It is known as the Great Barringer Meteor Crater.

1908 A mysterious explosion in the Tunguska region of Siberia flattens more than a million trees. The energy released is equal to that of twenty large hydrogen bombs. But no meteor fragments were found and the cause has never been conclusively determined. It is now believed that a celestial wanderer—an icy chunk of a comet or stony asteroid, perhaps a hundred fifty feet in diameter—exploded in the atmosphere approximately five miles above Siberia. (See Chapter 6, "Did an Asteroid Kill the Dinosaurs?," page 279.)

1909 French aviator-engineer Louis Blériot (1872–1936) makes the first flight across the English Channel, completing the crossing in thirty-seven minutes.

1909 American explorer Robert Edwin Peary (1856–1920) reaches the North Pole. He is accompanied by a team of Inuit and his personal assistant, Matthew Henson, a black explorer who is indispensable in dealing with the Inuit and overseeing the sleds and dog teams that will allow completion of the daring journey. Peary's claim was immediately disputed by Frederick Cook, a rival who said he had made it to the North Pole first, setting off a public-relations war between the two explorers. Most modern authorities, including the National Geographic Society, accept Peary's evidence and dismiss Cook's. Peary's book about his adventure, *Northward Over the Great Ice*, was sprinkled with nude photographs of his fourteen-year-old Inuit mistress and other Inuit girls, which Peary passed off as "ethnographic studies."

1911 Norwegian explorer Roald Amundsen (1872–1928) reaches the South Pole, beating the ill-fated Scott expedition by one month. Amundsen died when his plane was lost over the Barents Sea during a search for another explorer.

Geographic Voices   From the diary of Captain Robert Scott, found by searchers in November 1912

Sunday, 18 March. Today, lunch, we are 21 miles from the depot. Ill fortune presses, but better may come. We have had more wind and drift from ahead yesterday; had to stop marching; wind N.W., force 4, temp. –35°. No human being could face it, and we are worn out *nearly.* . . .

Thursday, 29 March. Since the 21st we have had a continuous gale from W.S.W. and S.W. We had fuel to make two cups of tea apiece and bare food for two days on the 20th. Every day we have been ready to start for our depot *11 miles away,* but outside the door of the tent it remains a scene of whirling drift. I do not think we can hope for any better things now. We shall stick it out to the end, but we are getting weaker, of course, and the end cannot be far. It seems a pity, but I do not think I can write more.

R. Scott

For God's sake look after our people.

1912 German geologist Alfred Wegener (1880–1930) publishes his theory of continental drift; it is dismissed until evidence collected since the 1960s shows him to be generally correct.

1912 The *Titanic,* a British passenger liner supposed to be unsinkable, hits an iceberg and sinks, leaving a death toll of some 1,500. (The wreck of the *Titanic* was found in 1985 off Newfoundland.)

1913 The Greenwich meridian is accepted internationally as the prime meridian.

1913 The auto-assembly line is introduced by Henry Ford. Cars are built as they move along a conveyor, reducing assembly time from twelve and a half, to one and a half hours, radically reducing the cost of an automobile.

1913 The Balkan Wars. In two wars, the Ottoman Empire loses almost all of its European territory.

1914 The Panama Canal opens, linking the Pacific to the Atlantic by way of the Caribbean Sea. The forty-mile canal was begun in 1903 after American warships aided a revolt that created the nation of Panama out of Colombia.

1914 Red and green traffic lights are introduced in Cleveland, Ohio.

1914 The First World War begins. It will last until 1918 and result in a redrawing of the maps of Europe and the various European colonial possessions around the world.

1915 German Hugo Junkers constructs the first fighter airplane.

1917 The Russian Revolution overthrows Tsar Nicholas II and installs a Communist regime under Vladimir Ilyich Lenin (1870–1924), founder of the Bolshevik Party and the Soviet state.

1917 The Trans-Siberian Railroad, the world's longest railroad, is completed. Begun in 1891, it stretches 5,787 miles from Moscow to Vladivostok on the Sea of Japan, and opens up the vast wilderness of Siberia to development.

1917 The Balfour Declaration promises a Jewish homeland in Palestine.

1919 The Treaty of Versailles, ending World War I, is imposed on Germany and its allies. Under the treaty, Germany is stripped of its African colonies and the Alsace-Lorraine region is given to France. At the same time, major portions of the Middle East—formerly part of the Ottoman Empire, which had been allied to Germany—were divided among the victorious allies. The British took control of Palestine, the Transjordan, and Mesopotamia. The French gained control of Lebanon and Syria. The former empire of Austria-Hungary and the states of Montenegro and Serbia disappeared from the map. They were carved up and would eventually become Poland, Czechoslovakia, and Yugoslavia. For the new Poland, a corridor to the Baltic Sea was carved out of Germany and the former German port of Danzig become the Polish city of Gdansk.

1920 Ireland is divided between predominantly Protestant Northern Ireland and the mostly Catholic Irish Free State. (Ireland became a Republic in 1949, severing all ties with England, but Northern Ireland remained a part of the United Kingdom.)

1922 The Union of Soviet Socialist Republics (USSR) is proclaimed. Lenin dies in 1924 and is succeeded by Joseph Stalin (1879–1953).

1922 The modern Republic of Turkey is formed out of the old Ottoman Empire. In 1930, the name Constantinople is changed to Istanbul.

1922 Iraq, formerly called Mesopotamia, is recognized as a kingdom.

1925 In South Africa, surgeon and amateur anthropologist Raymond Dart is given the skull of the "Taung child." Concluding it is neither human nor ape, Dart names the species *Australopithecus africanus*, or "southern ape from Africa." But the scientific community doesn't accept Dart's notion until 1950. Recent dating techniques found *A. africanus* to be 2.5 million years old, much older than Dart himself had thought.

1925 The German oceanographic ship *Meteor*, using recently invented sonar (*so*und *na*vigation *r*anging) discovers the Mid-Atlantic Ridge, the mountain range that lies under the Atlantic Ocean.

1925 The world's first motel, the Milestone Motel, opens in Monterey, California.

1926 American aviator and explorer Richard E. Byrd (1888–1957), later a naval rear admiral, becomes the first person to fly over the North Pole.

1927 American aviator Charles Lindbergh (1902–74) makes the first solo nonstop transatlantic crossing in *The Spirit of St. Louis*, completing the trip from New York to Paris in thirty-three and a half hours.

1929 The kingdom of Serbs, Croats, and Slovenes, carved out of the former Ottoman and Austro-Hungarian empires, is transformed into the state of Yugoslavia.

1931 A Spanish Republic is declared after the overthrow of the monarchy.

1933 In the United States, Franklin D. Roosevelt (1882–1945) is inaugurated for the first of four terms as president, while in Germany, Adolf Hitler (1889–1945) is granted dictatorial powers.

1935 The Nazis repudiate the Versailles Treaty.

1935 The name Persia is changed to Iran.

1936 The Germans occupy the Rhineland.

1937 American flier Amelia Earhart and her navigator are lost over the Pacific as they attempt to complete a flight around the world.

1939 Pan American Airways institutes the first regular commercial flights across the Atlantic.

1939 World War II commences, lasting until 1945.

1940 Following the secret nonaggression pact between Stalin and Hitler, the three Baltic nations of Latvia, Lithuania, and Estonia are annexed by the Soviet Union.

1940 In a cave in Lascaux, near Montignac, France, wall paintings dated to seventeen thousand years ago are discovered.

1943 Jacques Cousteau coinvents the Aqua-Lung, now better known under its acronym, scuba (self-contained underwater breathing apparatus).

1945 The United Nations is founded at a conference in San Francisco.

1947 The Dead Sea Scrolls are discovered.

1947 India is proclaimed an independent republic; it is then partitioned into predominantly Hindu India and predominantly Muslim Pakistan.

1948 The state of Israel gains independence.

1948 Burma and Ceylon are granted independence. (In 1972, Ceylon's name reverts to Sri Lanka, meaning "resplendent island.")

1949 The Communist People's Republic of China is proclaimed under the leadership of Chairman Mao Tse-tung (1893–1976).

1949 The North Atlantic Treaty Organization—NATO—is formed to protect West European nations from the threat of Communist attack by Eastern Europe. Its charter members are Belgium, Canada, Denmark, France, Great Britain, Iceland, Italy, Luxembourg, the Netherlands, Norway, Portugal, and the United States. NATO is later joined by Greece and Turkey (1951), West Germany (1955), and Spain (1982). (In 1966, France expels NATO forces and quits the organization.)

1949 The German Democratic Republic (East Germany) and the German Federal Republic (West Germany) are established, splitting Germany into East and West.

Why Did Hannibal Take Elephants Across the Alps, and Did Napoleon Know How Far It Was to Moscow?

"War," said the Duke of Wellington (1769–1852), a British general, "consists in getting at what is on the other side of the hill."

That is as simple a way as any of saying that wars are fought over geography. A key harbor, such as Port Arthur, controls shipping in or out of a country or a continent. Control the harbor and you control the ocean. Or a narrow pass through a mountain range becomes a funnel through which all men and materials must pass. The twenty-five-mile-long Khyber Pass, in the rugged mountains between Afghanistan and Pakistan, was historically the key access to India. It has been fought over for centuries since the Persian Empire and Alexander the Great. A town like Hastings in England might be the key to the lone road that leads to the heart of a nation. Another small, seemingly unimportant town, such as Gettysburg or Ypres, becomes a critical crossroads because of the roads that run through it. Generals of great armies have to feed thousands. Matters of food and supplies become complicated when invading forces, far from home and a reliable source of support, look to a supply depot or a rich food-growing area, which becomes a target of great significance.

Sometimes the geography is merely symbolic. A capital city, like Jerusalem, Washington, DC, or Paris, becomes a prize of great value. Or perhaps it is a fortified city like Verdun, or an outpost like Dien Bien Phu—a point on the map elevated into a crucible that will determine the turning of a war.

These matters of geography often do determine the outcome of wars—or at least of the battles that determine the outcome of wars. One vivid example of that came in one of the most significant battles in American history, the three days of bloody fighting at that small crossroads in Pennsylvania farm country called Gettysburg. As John Noble Wilford recounts in his book *The Mapmakers*, "it was at Gettysburg that [Gouverneur Kemble] Warren, chief engineer of the Army of the Potomac, cast his topographer's eye across the battlefield and recognized the strategic importance of Little Round Top. He led troops to seize the hill before the Confederates could, an action that proved decisive to the Union victory."

While the heroics of this soldier-mapmaker, immortalized in a plaque at the Gettysburg battlefield, may be a unique case of a cartographer having such an obvious impact on the course of a battle, there is no question that geography has had a great deal to do with determining where battles have been fought and, likewise, their outcome. Throughout military history, wise generals—and admirals—have always chosen their spots well, using terrain to determine tactics. The following is a list of some of the world's most notable battlefields; the battles fought there were all turning points in which geographical factors influenced the course of history.

World Battlefields That Shaped History

Marathon, Thermopylae, and Salamis Five hundred years before Christ, Persia was by far the world's largest empire, spreading from its base in present-day Iran over Mesopotamia, Asia Minor, Egypt, and Afghanistan. Greece was a collection of fledgling city-states, with Athens gradually emerging as the central power. When the Ionians— Greeks living in what is now Turkey—revolted against their Persian rulers, other Greek states stepped into what became known as the Persian Wars, lasting nearly fifty years.

Marathon was on an open plain in Greece, northeast of Athens. There, in 490 BC, an Athenian army under Miltiades, a Greek general who had once served under the Persian king Darius, defeated a Persian invading force twice its size. Fearing that Athens might surrender to a Persian fleet, Miltiades dispatched a runner, Pheidippides, to the city to report on the victory. On reaching Athens, he delivered the message, then collapsed and died. The Olympic games of ancient Greece commemorated this runner with the marathon, a race equal to the distance Pheidippides had covered.

Ten years after Marathon, the Persians, now led by Xerxes, again crossed into Greece with an estimated one hundred eighty thousand men. This army, perhaps at that time the largest ever seen in Europe, was met by a small force of three hundred Spartans under Leonidas at Thermopylae, a narrow pass in east-central Greece on the principal route from the north. Despite their enormous numerical superiority, the Persians were unable to break the Spartan hold on this strategic pass for three days. When the Spartans finally succumbed, the Persians captured and burned Athens.

But the war's decisive battle came at sea. Athens had built a large fleet that was massed in the narrow channel between Greece and the small island of Salamis, ten miles east of Athens. Once again, despite their numerical advantage, the Persians suffered a major defeat. The Athenians had earned an enormous victory that propelled Athens to the position of the central power of the Greek city-states. Although fighting between Greece and Persia would continue, Persia was essentially broken as a threat to Europe, allowing the flowering of the Greek period. In a much grander sense, it marked the beginnings of a European tradition that was distinct from Asia and would constitute the beginnings of Western civilization.

Cannae, Zama After the Greeks and Persians settled down, the next ancient superpower confrontation occurred between the Romans and their rivals for control of the Mediterranean, the Carthaginians. A coastal city near Tunis in present-day Tunisia, Carthage had been settled for hundreds of years by the Phoenicians, the great sailing empire based in what is now Lebanon. Three Punic Wars (*Punic* is Latin for "Phoenician") were fought between Rome and Carthage, lasting over one hundred years between 264 and 146 BC. In the first of these, Rome was able to evict Carthage from the island of Sicily.

From a base in Spain, then part of Carthage, the Carthaginian general Hannibal (247–152 BC) took a mixed force of North Africans, Spaniards, and Gauls and made a dramatic march, traveling with North African elephants (a breed now extinct) across the South of France and over the Alps in the winter of 218 BC. This audacious march, followed by quick victories in northern Italy, left Rome amazed and exposed to attack.

In need of supplies, Hannibal moved his army to Cannae, a major food depot on Italy's southeast coast, the center of Rome's corn-growing province. There, in 216 BC, Hannibal's army confronted the largest Roman army ever put in the field. But his tactic of drawing in the larger enemy and then encircling them, later called the "double envelopment," resulted in a humiliating loss for Rome and has since been studied and used by generations of soldiers. Despite his military success, Hannibal was unable to complete the victory over Rome. Trapped in the "toe" of Italy's "boot," he was eventually called back to Carthage when Roman forces attacked the city in 203 BC. Defeated by the Roman general Scipio at the battle of nearby Zama, Hannibal went into exile and eventually committed suicide.

In the third Punic War (146 BC), Rome finally defeated Carthage, reducing the city to a pile of rubble, plowing it under and enslaving the survivors. North Africa became part of the province of Africa in the expanding Roman Empire.

Hastings If, like most Americans, you received an "old school" education, 1066 is one of those dates you were supposed to lock into your brain. This is the year of the invasion of England by William the Conqueror (1027–87) of Normandy (France), a distant relative of the English king Edward the Confessor. William claimed the English throne that had been taken by Harold Godwin, the country's most powerful Anglo-Saxon nobleman, who had already turned back one threat to the throne by killing the king of Norway in battle.

William landed his force of knights and men-at-arms and marched to Hastings on England's southeast coast, where he quickly built a timber castle as a base. On October 14, the furious day-long battle, depicted in the famed Bayeux Tapestry, came to a close when the Norman archers rained arrows straight down on the Anglo-Saxons. In the following close combat, King Harold was hacked to death and the leaderless Anglo-Saxon forces disintegrated. On Christmas Day, 1066, William was crowned king of England in Westminster Abbey and began a century of Norman rule over England that would extend into much of Europe and play a large part in the Crusades to recapture the Holy Land.

Acre, Arsuf The Crusades began in 1095 after the earlier capture of Jerusalem by the Seljuk Turks brought a call from the pope to recover the Holy Lands of Palestine from the Muslims. The Europeans were initially successful in retaking Jerusalem. But it was taken back in 1187 by Saladin (1138–93), the Kurdish general who controlled much of North Africa and the Middle East. Known as a patron of the arts and renowned for his chivalry, Saladin was nonetheless a Muslim and considered a heathen by the Europeans, who wanted possession of the holy city of Jerusalem and the True Cross, purported to be a fragment of the cross Christ was crucified on and Christianity's most sacred relic.

In 1189, the Third Crusade was called and, after a false start, was led by England's King Richard I (the Lion-Hearted of Robin Hood fame;

1157–99), and France's Philip II, who happened to be at war between themselves at the time! They set aside differences long enough to take the town of Acre (Akko), a port on what is now Israel's northwestern coast. Holding some twenty-five hundred civilians hostage, Richard I put these captives to the sword in an appalling murder of women and children, not at all in line with Hollywood's sainted vision of "Good King Richard."

Richard then marched south along the coast toward the port of Jaffa (Yafo), shadowed by Saladin's troops, until the two armies reached Arsuf, a spot on the coast a few miles north of where Tel Aviv now stands. In intense heat, the two armies faced off, Richard's crusaders with their backs to the sea. Maintaining a disciplined defensive posture until he called for a counterattack, Richard was able to take control of the battle quickly and drove Saladin's army from the field.

Unable to recapture Jerusalem, Richard settled for a treaty with Saladin allowing Christian pilgrims free access to the city's shrines. The English king fared less well on the way home. He was captured by the king of Austria, who held him for ransom, which was ultimately raised by his English subjects.

Saratoga The early battles of the American Revolution in the state of Massachusetts at Lexington and Concord—with the famous "shot heard 'round the world"—or that of Bunker (actually Breed's) Hill in Boston may be more famous, but the series of battles fought between the rebellious Continental Army and the British troops in Saratoga, New York, in the fall of 1777 were far more significant in the war's eventual outcome.

Moving south from Canada, British troops were supposed to meet another force coming up the Hudson River from New York. The plan was to cut off New England, the center of colonial revolt, from the rest of the American colonies. But poor planning and communications doomed the British plan. The first of two pitched battles took place on September 19, 1777, at Saratoga, about 25 miles (40 km) north of Albany, New York, where about seven thousand Americans had dug in. Among them were five hundred sharpshooters from Pennsylvania under the command of Daniel Morgan, who had an easy time shooting from concealed positions against British regulars who were employed in the traditional European battle square. British officers in distinctive gold-braided uniforms also made easy targets for the American marksmen. The first battle ended inconclusively with the British taking heavy casualties but setting up earthwork defenses.

On October 7, the second battle of Saratoga commenced. Led by the flamboyant Benedict Arnold—later to turn traitor to the colonial cause—the Americans overwhelmed the British. Ten days later, British general Burgoyne surrendered, blunting a major British offensive against the colonial rebels and removing an eight-thousand-man British force from the picture. The immediate strategic impact of the victory left the American rebels in control of the Hudson River. More significantly, almost immediately after Saratoga, Americans and Englishmen began to feel for the first time that a rebel victory was possible. That sentiment was also expressed in France, Great Britain's chief rival, which allied itself with the American cause and ultimately turned the tide of war in favor of the rebellious colonists.

Trafalgar, Austerlitz, Borodino, and Waterloo Following the American victory over the British, continental Europe descended into a maelstrom of conflict, beginning with the French Revolution of 1789, which established the French Republic, only to be replaced quickly by the rise of Napoleon Bonaparte (1769–1821). After a series of victories over armies from Austria, Russia, and Prussia (later to become the chief part of a unified Germany) who were committed to restoring France's monarchy, Napoleon returned to France and engineered a coup that made him virtual ruler in 1799. In 1804, he was proclaimed emperor of France, with the blessing of the pope. But his ambitions went beyond France itself to a continental Europe dominated by France. He planned an invasion of England, but his overreaching scheme led to years of fighting, known as the Napoleonic Wars, that raged across the continent from 1804 to 1815.

The naval battle fought on October 21, 1805, off Cape Trafalgar on the southwest coast of Spain near the Strait of Gibraltar was the decisive naval engagement of the wars. After England's Admiral Nelson, who died in the battle, defeated Napoleon's French navy, British sea supremacy was established—and maintained for the next century.

Frustrated by this defeat in his scheme to attack England, his most implacable foe, Napoleon turned his sights and his army to the continent.

Marching from France into Central Europe, Napoleon confronted the allied armies of Tsar Alexander I of Russia and Emperor Francis I of Austria. The major battle came near the village of Austerlitz, near present-day Brno, a city in southern Czechoslovakia.

The fighting there in December 1805 has been called Napoleon's most tactically perfect battle. Although the allied army Napoleon faced was much larger, it suffered problems of communication—its leaders and troops spoke different languages. To make matters worse, the young tsar Alexander I took command of the troops from his experienced field marshal, Kutuzov. Napoleon drew the larger army away from its position on a strategic height, captured those same heights, and easily annihilated much of the army that had been sent to crush him. A day after the battle, the Austrians and Russians came to terms with Napoleon. During the next two years, Napoleon mastered most of continental Europe, installing family members on most of the thrones.

But he suffered his first setback on land in the Peninsular War fought for control of Spain and Portugal. The fighting there began in 1808 and dragged on for six years. And in 1812, facing defeat in Spain and Portugal, Napoleon made a disastrous decision to invade Russia. With half a million men, Napoleon left France to march on Moscow. Along the route, he had to leave behind troops to secure his gains. By the time he reached Russia, his army was smaller and a long way from home and secure supplies.

At Borodino, a village 70 miles (110 km) west of Moscow, he was met by the Russian Army under the command of Field Marshal Kutuzov, whose sound advice had been ignored by the tsar seven years earlier at Austerlitz. In a battle immortalized in Tolstoy's *War and Peace*, Napoleon broke through and cleared the way to Moscow. Although he occupied the Russian capital, he did so at great cost. The onset of the severe Russian winter took an enormous toll on his army. Napoleon was forced to leave Moscow and return to France, with Kutuzov in a punishing, vengeful pursuit. By the time they reached Paris, Napoleon's army of half a million was reduced to thirty thousand by the cold and the brutal fighting in Russia. In 1814, France was attacked by a new alliance of Russia, Austria, Prussia, and Great Britain. Unable to repel this invasion, Napoleon was forced to abdicate and was sent to the isle of Elba, off the coast of Italy, for the first of two exiles.

But even as the allies set about to restore the French monarchy, Napoleon escaped and returned to France. Welcomed like a returning messiah, he marched on Paris, gathering supporters along the way. Although Napoleon promised peace, his enemies abroad immediately declared war. Past his prime and in poor health, Napoleon gathered an army of 125,000 battle-tested veterans and set off for Belgium and an encounter with the British general Lord Wellington (Arthur Wellesley, 1769–1852).

On June 18, 1815, the armies met near Waterloo, a Belgian village south of Brussels. Fighting across acres of farm fields in a torrential rain that turned the land into a gory, muddy quagmire, Napoleon's last army was soundly defeated. Four days later, he abdicated again. Exiled a second time, he was sent to St. Helena, a desolate volcanic island in the South Atlantic, where the former emperor of France and master of Europe died in isolation.

Gettysburg Fought over the first three days in July 1863, the battle at Gettysburg, Pennsylvania, was the turning point in the long, bloody War Between the States, or Civil War. Knowing that his outnumbered troops could not win a long, drawn-out war against the richer, more populous, industrialized northern states, Confederate general Robert E. Lee (1807–70) attempted to carry the war to the North. After a series of victories against inferior Union commanders, Lee took his seventy-five thousand seasoned troops through the rich Shenandoah Valley of Virginia and into Pennsylvania. He hoped to outflank the Union Army, then drive south and take Washington, DC, demoralize the enemy, and quickly end the war with European recognition of the Confederate States.

More by accident than design, a party of Confederate scouts, said to be looking for shoes, encountered a patrol from the Union's Army of the Potomac. They met at the small town of Gettysburg, a crossroads for a railroad line and a dozen roads leading to all points of the compass. After this initial accidental engagement, commanders on both sides poured troops into the town, with the Union soldiers arriving first and in sufficient force to take the high ground that would prove to be of decisive importance in this long, ghastly battle. A brilliant commander who had reluctantly left the United States Army when his home state of Virginia seceded from the Union, Lee knew victory would clear the way for a swift strike at the nation's capital. He had the confidence of a general whose armies had outfought those of the Union at nearly every turn.

For three days, the fighting was fierce, often hand to hand. Against wave after wave of Confederate infantry charges, the Union forces held their ground until Lee was forced to withdraw from the field. Tens of thousands died on both sides, and Lee's army, dealt an almost fatal blow, limped back toward Virginia. The failure of the Union commander to pursue the bedraggled southerners and finish them off was a controversial decision that may have prolonged the war for several more years of bitter fighting, ultimately reducing the South to smoldering ruins.

With the Union victory, Confederate hopes for recognition and assistance from Europe withered overnight. Lee's best army, having lost its aura of invincibility, would never again be able to withstand the massive Union advantages in manpower, supplies, and wartime production.

Little Big Horn One of those present at Gettysburg was an ambitious, twenty-three-year-old "boy general" whose cavalry held off the Confederate reinforcements under one of Lee's ablest generals, Jeb Stuart. That Union cavalryman was George Armstrong Custer (1839–76). Hailed for his performance at Gettysburg, Custer would achieve a kind of immortality for his part in another bloody disaster, the Battle of the Little Big Horn, also called Custer's Last Stand. An eccentric egotist, Custer wore black velvet uniforms he had designed himself and shoulder-length blond hair; the Indians called him "Long Hair."

After witnessing the slaughter by whites of great herds of buffalo as the railroads moved west, the Plains Indians attempted a defiant stand to retain the lands they had been guaranteed under peace treaties made—and broken—by the federal government. When Custer led a band of prospectors into the Black Hills of South Dakota and found gold, prompting a gold rush for Indian lands, the scene was set for a showdown between the Plains Indians and the "pony soldiers."

Ignoring the advice of scouts who reported the large numbers of Indians gathering against him near the Little Bighorn River valley in Montana, Custer divided his command and ordered an attack. Without making proper reconnaissance, Custer was sure that his troopers could handle the thousand Indians he expected. On June 25, 1876, hopelessly outnumbered by more than three thousand Sioux warriors, his entire command was slaughtered in about half an hour of fighting.

Word of the battle, which reached back East as the nation was celebrating its centennial, prompted calls for massive retribution against the Indians. This battle, one of the few Indian victories in their three-hundred-year struggle against the inexorable European invasion, made Custer a revered American folk hero for years to come. Only recently has his foolhardiness, ego, and overreaching ambition been acknowledged as the true reason for the brutal end he met.

Port Arthur Closed to the West until the arrival in Tokyo Bay of Matthew C. Perry in 1853, Japan had remained far from involvement in European and American affairs. But the opening to the West in the second half of the nineteenth century brought rapid modernization to Japan and fueled that country's ambitious plans to gain a piece of the great Euro-American colonial expansion into Asia. By the turn of the century, the heretofore reclusive Japanese were ready to become players on the world stage. With the once-formidable Chinese empire in disarray as European countries carved off parts of China and claimed them as "spheres of influence," Japan first set its sights on controlling Korea and Manchuria, a goal that brought a head-on collision with Russia in the Russo-Japanese War of 1904–5.

Japan's first military target was the Russian-held Port Arthur, a strategic deepwater harbor on the tip of the Liaodong Peninsula in what is now China. It was then Russia's only ice-free harbor on the Pacific; control of the port meant command of the local seas. The Japanese attack on the fortified city, in which the full fury of a twentieth-century militarized industrial state was brought to bear, lasted from mid-August 1904 to January 1905. When the Russian commander finally surrendered, the Japanese had suffered sixty thousand casualties; the Russians numbered thirty thousand dead and wounded.

The fall of Port Arthur forced the Russians to sue for peace, which was negotiated by President Theodore Roosevelt in Portsmouth, New Hampshire. Russia ceded Port Arthur and part of Sakhalin Island, and was forced to leave Manchuria. Japan was left as the predominant power in East Asia and would seek to expand its military power and influence in the Pacific during the next three decades, a policy that eventually produced a collision with the United States in the Second World War.

Marne, Tannenberg, Ypres, Gallipoli, Verdun, and the Somme When the First World War broke out in the summer of 1914, Germany's plan was simple: drive swiftly through Belgium and deliver a knockout blow to France. Then Germany would turn its attention to the east and contend with Russia. The first part of that sweep brought war to the low, flat plains of Belgium and central France, the scene of fighting a century earlier in the Napoleonic Wars.

The German strategy, known as the Schlieffen plan, was initially successful as the kaiser's armies cut a wide swath through Belgium and bore down on Paris until they met an Allied army on the banks of the Marne, a river in central France just north of Paris. The first of two battles was fought there in September 1914. After suffering this first defeat, the Germans withdrew and their initial impetus stalled. When both sides dug in, the long, terrible stalemate of World War I's trench warfare began.

A few weeks earlier, the Germans had exacted a severe toll from the Russians, their opposition on the eastern front. In fighting near Tannenberg, a small village in what is now Poland, the Germans turned back a Russian invasion intended to weaken Germany by creating battlefronts on two sides. But the Russian troops were badly commanded, poorly equipped, and ill trained; they met disaster when they faced the modernized, cohesive German army. After thousands of starving Russian soldiers surrendered, the Russian commander committed suicide. The war on this eastern front continued inconclusively until 1917, when the Russian Revolution overturned the tsar and Lenin's Bolshevik government withdrew from the war.

Back on the western front, the Germans concocted a plan to cut across Belgium and take the port cities on the English Channel. One of their targets was the small lace-making town of Ypres, a communications hub and strategic crossroads. Over the next four years, three battles would be fought over Ypres, producing a million casualties on both sides. Many of the dead are buried in the forty cemeteries that dot the vicinity and which would eventually give the town its grisly nickname, Ypres la Morte. The first battle, in October 1914, began the long period of trench warfare that so senselessly chewed up the young men of Europe.

By the spring of 1915, the war in Europe had turned into a terrible and costly stalemate. Fierce nationalistic pride and the certainty of having superior armies kept both sides from negotiating a settlement. A British plan to break the deadlock called for a new offensive through the Ottoman Empire (modern Turkey), then allied to Germany. The point of attack was the narrow Gallipoli Peninsula overlooking the Dardanelles, the strait that connects the Mediterranean to the Black Sea. The plan had merit, but the British failed to launch it in time. By the time they did attack, the Turks and Germans had heavily fortified the area in anticipation. The combined assault by British, French, Australian, and New Zealand troops began in August 1915. Instead of a brilliant strategic coup, the assault on Gallipoli turned into a fiasco, opening up one more long, ghastly front in which losses on both sides were enormous and without the British and French gaining the advantage they'd hoped for.

The stalemate on the western front continued into 1916. Both sides mounted ambitious offensives to put an end to the war. The Germans assaulted the fortress city of Verdun, on the Meuse River in eastern France, in February 1916. The fighting raged inconclusively but devastatingly for months. By June, the French had lost more than 300,000 men at Verdun, the Germans 281,000.

Greater madness and death lay ahead when the French and Allies counterattacked in July 1916 in the Battle of the Somme, a river in the north of France. The fighting there lasted until November, and when it was over, the conflict had claimed over a million casualties at Somme alone.

The aftermath of the Somme was simply another deadly but indecisive standoff. When the Germans attempted a last-ditch offensive, once again at the Marne, the allies fended them off. Exhaustion rather than good sense finally prevailed. The United States had joined the British and French, providing fresh cannon fodder for a time, but by 1918, the Allies had at last begun to advance into German territory, and Germany and its allies called for peace.

Stalingrad, El Alamein, Normandy, and Iwo Jima The First World War had been called the Great War and even the "war to end all wars." That was before the world realized that a sequel was imminent.

With Germany crippled by its loss, having surrendered both European territory and colonial possessions overseas, there was great unrest in the country. During the economic upheaval of the worldwide depression of the 1930s, Adolf Hitler (1889–1945) came to power in Germany and blamed the country's problems on the unfair war reparations that Germany had to pay, on foreigners in general, the French in particular, and on the Jews, Hitler's most tragic scapegoats. He found an accepting national audience eager to see Germany restored to its "rightful" place among the nations of Europe.

After making quick work of Czechoslovakia and Poland in 1939, Hitler's armies smashed through Denmark and Norway in the spring of 1940. And in a repeat of the strategy of 1914, Germany rolled over the Low Countries of Belgium, the Netherlands, and Luxembourg and then into France, taking Paris against token resistance. The British made a heroic moment of their evacuation from the French port of Dunkirk. Hitler's Germany and Mussolini's Italy had uncontested control over Europe, and the German armies once again wore the sash of seeming invincibility.

A secret nonaggression pact with the Soviet leader Stalin temporarily kept the Russians out of the war and allowed Stalin to take half of Poland and the three Baltic states of Lithuania, Estonia, and Latvia. But thwarted by the British victory in the Battle of Britain—the monumental air war fought in the skies over Great Britain—Hitler made the same mistake Napoleon and the German commanders of 1914 had made: he invaded Russia late in 1941. These Germans, too, learned the hard lesson of committing land troops to Russia. The German offensive stalled at the monumental Battle of Stalingrad, an industrial center (formerly called Tsaritsyn, and renamed Volgograd in 1961) at the junction of the Don and the Volga Rivers. Fought from August 1942 to early in 1943, with both sides suffering heavy casualties, the Russian victory there marked the end of Germany's eastern offensive.

The United States entered the war after the attack on the naval base at Pearl Harbor by Germany's ally Japan in December 1941, and the American-British Allies planned to counterattack. But instead of assaulting Europe directly, they planned first to retake North Africa. The Italians had moved on Libya and Algeria in 1940 in an attempt to control the

Mediterranean and take Egypt, the Suez Canal, and the Middle East. The obvious prize was the oil, without which the modern military machine could not run. But Mussolini's troops failed to complete this conquest and were thrown out of Egypt and eastern Libya. The Germans then moved into North Africa under the command of one of its most able generals, Field Marshal Erwin Rommel (1891–1944).

Fought from late October to November 1942, the Battle of El Alamein was one of the turning points of the Second World War. El Alamein was a small railway junction located west of Alexandria, Egypt, in the desert on the Mediterranean coast. The British, under Bernard Montgomery (1887–1976), held a line of defense that stretched 40 miles (64 km) from El Alamein into the desert. To attempt to take the Suez Canal, Rommel would have to pass this line. But Rommel's attack was pushed back and Montgomery then went on the offensive in an attempt to force the Germans west, where the newly arrived Americans, under Dwight Eisenhower and George Patton, were landing in Algeria and Morocco. With this pincer action, the Allies hoped to force the Germans and Italians out of North Africa and prepare for an invasion of Europe through Italy. Twelve days of vicious fighting across heavily mined desert wasteland laced with barbed wire ended with a British victory; it was their first in a land battle over the previously invincible Germans, who were forced to withdraw from North Africa. The Middle East's oil was safe and the Allies had a staging area for their next campaign, the 1943 assault on Sicily, the Mediterranean island whose possession would allow naval control of the sea and an invasion of Italy to follow in 1944.

Two days after Rome fell to the Allies, the largest invasion force in history assaulted the beaches of Normandy on the coast of France's English Channel between the port cities of Cherbourg and Le Havre. On June 6, 1944—D-Day—4,000 invasion ships, 600 warships, 10,000 planes, and more than 175,000 troops were committed to the assault. After four days, the Allies had secured the beachhead from which they would launch the counteroffensive against Germany. Nearly a year of intense fighting across France, Belgium, and finally into Germany itself followed before Berlin fell in May 1945.

After the disaster at Pearl Harbor, the United States had committed most of its forces to the European theater. But fighting in the Pacific against Japan continued, with the Japanese military machine virtually unchecked until well into 1942. By that time, Japan's forces controlled nearly 10 percent of the earth's surface.

When the Americans finally took the offensive in the summer of 1942, it marked the beginning of a grueling, grinding, hard-fought campaign to retake a series of small Pacific islands that would provide stepping-stones for the eventual planned invasion of Japan. The first of these islands was Guadalcanal, and the huge losses taken there by both sides were a grim indicator of the horrific warfare that would continue in the Pacific for the next few years. The succession of islands—Tarawa, Saipan, Guam, Tinian, and the Philippines—saw desperate battles between beach-landing U.S. Marines and Japanese defenses that had been prepared for years.

The deadly progression through the Pacific culminated with the invasion of Iwo Jima, an eight-square-mile piece of volcanic rock dominated by an extinct volcano, Mount Suribachi. A well-protected but critical obstacle in the way of American operations, Iwo Jima was home to a Japanese airfield that both alerted Japan to impending American air raids and sent up fighters to attack Allied bombers on their way to the main islands. If taken, the Japanese airfield would provide an airstrip from which American fighters could protect bombers and where any crippled American planes could safely land after striking Japan.

The island was bombed around the clock for three months before the Marines were sent in. But the Japanese were well protected in a honeycomb of concrete bunkers and well-stocked tunnels beneath the island's hard volcanic surface. It took weeks of intense fighting to finally take control of Iwo Jima, an event immortalized in a photograph of the American flag being raised on Mount Suribachi (a staged, not spontaneous, event, according to recent revelations). The Marines suffered nearly 7,000 ,dead and another 15,000 wounded; only 1,000 of the 21,000 Japanese defenders survived. One more bloody assault on Okinawa, a large island south of Japan, lay ahead. With its fall in June 1945, Japan was cut off and reeling under the constant pounding of American bombers, which now attacked with near impunity. The firebombing of Tokyo in March 1945, for instance, killed 100,000 and left Tokyo in flames. A few months later, in August 1945, the war came to a close with the dropping of atomic bombs on the Japanese cities of Hiroshima and Nagasaki, the justification for which is still debated today.

Dien Bien Phu With the end of the war against Japan, Europe and America moved quickly to assert their control over Japan's Asian holdings. The French tried to reestablish control over old colonies in Southeast Asia—Laos, Cambodia, and Vietnam. But the Vietnamese, under Communist leader Ho Chi Minh and his brilliant general Vo Nguyen Giap, balked and expected independence. War came in 1946, with the United States openly aiding the French effort to maintain control of the country. A largely guerrilla war was fought for the next seven years, with the French controlling the cities but the Viet Minh guerrillas holding the countryside. Struggling against a largely invisible enemy, the French command hoped to lure the Viet Minh into a pitched battle and chose the hamlet of Dien Bien Phu, a small village 200 miles (320 km) west of Hanoi, as the site of the trap.

But the tables were turned. The Viet Minh began a long, deadly siege of the French fortifications at Dien Bien Phu. As the Western world watched, the French garrison was depleted in the six-month siege, and the United States contemplated military assistance. The possibility of using an atomic weapon against the Viet Minh was even proposed by President Eisenhower. In May 1954, the French succumbed and surrendered. A peace treaty followed, dividing Vietnam into two separate states, the Communist north and the anti-Communist republic in the south.

Having learned nothing from the disastrous French experience and fearing a "domino effect" on the rest of the region if South Vietnam also fell to the Communists, the United States almost immediately replaced the French in providing aid and then troops to South Vietnam in the ultimately disastrous American involvement in Vietnam during the 1960s and early 1970s.

Country, Nation, Republic, State: A Geopolitical Primer

Ever since George H. W. Bush proclaimed a "new world order," there has been more chaos around the globe than the tidy scenes of blissful cooperation the American president wistfully envisioned. In spite of

Bush's proclamation, little about the affairs of the world has been new, and much less has been orderly. In the wake of the demise of the Soviet Union, there have been bloody civil wars in several former Soviet republics. And Yugoslavia—arbitrarily carved out of the remains of the Austro-Hungarian Empire after World War I—has been shattered by a brutal outburst of ethnic and nationalistic fighting that has left the country broken into pieces. Apparently, there is no oil in Sarajevo, however. So the "coalition" that "liberated" Kuwait wasn't equally inspired to heroic deeds by the sight of merciless bloodshed in the Balkans.

Apart from the tragedy of Yugoslavia, the extraordinary events in Europe during recent times have clearly transformed the world that was left after World War II. For nearly half a century of cold war, two contending camps led by the United States and the Soviet Union struggled for supremacy. Every local brushfire war around the globe was fanned into a superpower crisis. And Cuba, Vietnam, Korea, the Middle East, and Nicaragua were regional conflicts that threatened to burst into much larger conflagrations, always with the specter of a mushroom cloud lingering in the background. Ironically, the civil wars in Yugoslavia and the former Soviet republics have largely been ignored by the United States and Russia. A decade or two ago, these conflicts would have had tanks rolling across borders, tens of thousands of troops on alert, and twitchy fingers in Moscow and Washington edging closer to war buttons.

But now the Union of Soviet Socialist Republics is no more. Shrunken by the shedding of its former *republics*, the former USSR is now the Commonwealth of Independent States. Every day, new *nations* apply for membership in the United Nations. In a halting yet undeniably historic series of Middle East peace talks—unthinkable only a few years ago—the Palestinians who were displaced by the establishment of Israel press for an independent homeland. Meanwhile, that's exactly what the government of Canada has authorized for the native Inuit (as they call themselves) or Eskimos (as they are more commonly known). *Nunavut*, an Inuit word for "our land," will become a self-governing homeland, carved out of a huge area that was once Canada's Northwest Territories.

Every day newspaper headlines are filled with geographic political terms that probably leave a good many people confused as to what exactly the difference is—if any—between a nation, a country, and a state. The following glossary is designed to sort out, simplify, and clarify the meanings of some of these commonplace, but misused and abused, terms.

Colony A word drawn from Latin meaning "to cultivate," as in a garden. Colonies were originally thought of as settlements of people transplanted from one place to a new, often undeveloped or sparsely populated area. For example, the Vikings who came to North America around the year 1000, the Puritans who established the Massachusetts Bay Colony, and the first waves of British sent to populate Australia were establishing true colonies. If Earthlings ever get to Mars and settle there, that will be a colony.

While that original meaning of colony still holds true, another sense of the word has taken hold. In time, colonies came to mean the territory conquered and controlled by a distant, and usually stronger, power. At its peak, for instance, the British Empire held colonies on nearly every continent, including the whole of India, and large sections of Africa and the Middle East. Although many British diplomats, soldiers, and civil servants were sent to govern in these colonies, they were always a minority in place to control a large native population rather than settlers dispatched to cultivate largely unsettled territory.

Commonwealth Not too long ago, this word meant only one thing: the British Commonwealth, which officially became the polite way of referring to the British Empire in 1931. The sun finally set on the empire after the devastation of World War II and the loss of its extensive overseas possessions during the independence movements of the 1960s. But the British attempted to keep its once mighty empire intact through the British Commonwealth, if only with looser strings, as an economic and trade partnership. Nowadays it is called the Commonwealth of Nations, a partnership of many of the former British colonies, which are now mostly self-governing but have retained a special economic relationship with the United Kingdom.

The word *commonwealth* literally means "communal well-being," and is used by a variety of political groupings around the globe. Australia, the Bahamas, and the island of Dominica are each formally called a commonwealth, as are several American states and the territory of Puerto Rico. The latest commonwealth is the one formed out of the remains of the Soviet Union. The Commonwealth of Independent States

is a group whose function is not yet clearly defined and whose future seems hazy at best.

While a strict definition of a commonwealth might be somewhat fuzzy, it boils down to a sense of union based on mutual interests.

Confederation Switzerland is the lone nation that calls itself a confederation, which generally refers to a group of states united for a common purpose.

In the wake of the American Revolution, the new nation of America was briefly, and not very successfully, governed under the Articles of Confederation, which provided for a weak central government. When that failed, the thirteen former colonies got together and wrote the American Constitution. A second American confederation came about with the secession of the southern states into a confederacy that led to the outbreak of the Civil War in 1860.

A federal *republic*, the United States of America is in essence a confederation or collection of states that retain certain powers and autonomy that cannot be overruled by the central government.

Country In strict usage, this is a geographic term signifying the physical territory of one nation, but it is often used in the extended sense of *nation*, regardless of whether it is dependent or independent. So while it is typical to ask "What country do you come from?" what we really mean is "What is your nationality?"

Coup d'état (sometimes just plain *coup*) This French phrase translates as a "stroke of the state," and means the sudden violent overthrow of a government. If it's not so violent, it's often called a "bloodless" coup. But that is a very relative term, depending on whose ox has been gored and whose blood has been spilled.

Imperialism For years, Communists around the world derided Americans as "imperialists." And when President Ronald Reagan wanted to say exactly what he thought of the Soviet Union, he minced no words, calling it the "Evil Empire."

An overused and casually tossed term of insult, *imperialism* means the acquisition or control of territory by a state in order to exploit the resources of that territory. The reason is usually simple: the controlling nation exploits the raw materials of a colony or territory, produces finished goods, and then sells them at a great profit, often back to the colony from which the materials came. Perhaps the greatest historical

example of the imperial system was Great Britain in India, wherein a small imperial nation controlled a vast colony. Cotton grown in India was shipped back to England, where it was converted to cloth in the mills of Manchester, sewn into clothing, and then shipped back to India for sale. The cost of the finished goods being much higher than the raw material, England always profited from this trade.

The Indian nationalist leader Mohandas Gandhi (1869–1948) understood this relationship perfectly and used it to formulate a simple but effective protest. He implored Indians to shun foreign-made cloth and begin wearing "homespun," or clothing that they had made themselves. This elementary and nonviolent tactic proved to be both a symbolic gesture of independence and an enormous economic blow to the British. It was one of the most effective of the many tools Gandhi used in his peaceful quest for Indian independence, which was finally granted in 1947.

Nation Generally, *nation* denotes a relatively large group of people who want to be organized under a single, usually independent, government. This group's members are often closely associated by common cultural characteristics—that is, they share common origins, history, and frequently language, as well as customs, values, and aspirations. Conscious of their own sense of nationality, such a group wishes to remain free from outside political domination.

Nation-states, such as those that comprise the membership of the UN, can be variously described as republics, principalities, kingdoms, or commonwealths.

But there is an alternative definition of *nation*—a federation or tribe. This is true when referring, for instance, to groups of North American Indians, such as the Cherokee or Sioux nations.

There are also nations without *states*. The 21 million Kurds who are spread out over large parts of Turkey, Iraq, and Iran, and spill into Syria and the former Soviet Republic of Armenia, constitute the world's largest nation without a territory to call their own state. Suppressed by the Turkish government and gassed by the Iraqi regime of Saddam Hussein without any protest from Western nations, the Kurds remain a volatile factor in an already tense part of the globe. With an ancient history of tribal conflicts and lacking a single Kurdish language, the Kurds have been internally divided by their own history and their geography.

Mountains within the area known as Kurdistan limit the ability of the Kurds to coalesce in a single nationalist movement.

Kingdom Lands ruled by kings somehow seem like vestiges of fairy-tale times. But there are still quite a few nations that are led by kings, either in name or in fact. The list of existing kingdoms includes Belgium, Bhutan, Denmark, Thailand, Jordan, Lesotho, Morocco, Nepal, the Netherlands, Norway, Saudi Arabia, Spain, Sweden, Swaziland, Tonga, and, of course, the United Kingdom, even though the current monarch, Queen Elizabeth, is not a king. Nearly all of these kingdoms are constitutional monarchies in which the figurehead king or queen's actual powers are largely symbolic and political power rests with an elected body such as the British Parliament. But in several of them, notably Jordan and Saudi Arabia, the king retains substantial, if not total, ruling power.

Principality This one is easy enough. It's like a kingdom, except that it's ruled by a prince. The three existing principalities are the small European states of Andorra, Liechtenstein, and Monaco. All three are constitutional monarchies in which elected governments actually rule. In addition to kingdoms and principalities, nations that are headed by some sort of royal family include the Sultanates of Oman and Brunei, the Grand Duchy of Luxembourg, the Emirate of Kuwait and the United Arab Emirates.

Republic Taken from the Latin meaning "thing of the people," this is the word most independent nations apply to themselves today. In a general sense, *republic* means a political entity whose head of state is not a monarch and, in modern times, is usually a president. Theoretically, the supreme power in a republic lies with a body of citizens who are entitled to vote for representatives responsible to them. Theoretically, that is, since there are any number of "people's republics" in which the people have little or no say in the appointment of officials or the course of their government.

State This is really one of the more confusing terms because it can go two ways. *State* can be applied to an entire nation, as in the "state of Israel" or the "Eastern European states," or to one of the territories and political units comprising a federation under a sovereign government, as in the United States of America or the various states of India. In the first sense, a state is the equivalent of a nation, an area of land with clearly defined boundaries and which has internationally recognized legal independence.

Afghanistan to Zimbabwe: How Many Nations Are There in the World?

With the shake-up of Eastern Europe, the demise of the Soviet Union, and numerous other assorted changes to the world map, there are now 178 member countries in the United Nations. (See Appendix III: The Nations of the World.)

But don't put your calculators away. The pace of country-making hasn't slowed since 1992. The new republics of Croatia, Slovenia, Bosnia and Herzegovina, and Macedonia have already splintered off from the former Yugoslavia, which also split into Serbia and Montenegro.

This total also does not include Greenland, a fairly significant piece of real estate that is an autonomous country within the Kingdom of Denmark. Also not included in this count are the four so-called republics of Bophuthatswana, Ciskei, Transkei, and Venda. These are tribal "homelands" that lie within the boundaries of South Africa but have only been recognized by South Africa as independent republics. For all practical purposes, these homelands are still governed by South Africa.

And if returning land to rightful ownership was the point of the Gulf War against Iraq, why stop at Kuwait? Since much decision-making power in the UN Security Council rests in the hands of the "Big Five" permanent members, start with their record of invasions, annexations, incursions, and general misbehavior.

The United Kingdom Few nations were more vocally outraged over the invasion of Kuwait than the British. As Parliament resolutely glared at Iraq, it may have been easy for them to ignore their own backyard. Northern Ireland hasn't gone away, and British troops still occupy a piece of land that is obviously attached to another country. While the rest of Ireland gained independence in 1922, the British retain this vestigial bit of colony based on a centuries-old conquest. The British also hold several other jewels, baubles from its shattered imperial crown, including Gibraltar, the Virgin Islands, the Cayman Islands, Bermuda, and the Falkland Islands, a desolate piece of real estate lying off the coast of Argentina, which attempted to retake them in a war in 1982. At least Hong Kong was returned to Chinese rule in 1997.

France Hand it to the French. They cannot be accused of bad taste in their choice of territorial possessions. Guadeloupe and Martinique are lovely Caribbean paradises that remain in French hands. A group of Pacific islands comprises French Polynesia. The French also retain French Guiana, the last European possession in South America, where France's satellites are launched, and New Caledonia, a major source of nickel.

The United States of America In chronological order, begin with Texas and much of the Southwest and California, "purchased" by treaty after the Mexican War from 1846 to 1848, a war described by Ulysses S. Grant as one of the "most unjust ever waged by a stronger against a weaker nation." (Such a purchase is the diplomatic equivalent of godfather Don Corleone's "offer they can't refuse.") Half a century later in the aftermath of the Spanish-American War in 1898 came the acquisition of Puerto Rico, today a U.S. commonwealth. But unlike the Commonwealth of Massachusetts, Puerto Rico's residents do not get to vote for president and enjoy no congressional representation. While the issue of Puerto Rican independence creeps toward a resolution, the question remains in the hands of the U.S. Senate rather than with the Puerto Rican people. Also in the Caribbean, the United States retains its Virgin Islands possessions and a base on Cuba at Guantánamo Bay. And way over in the South Pacific are a passel of lovely islands—all leftovers from the war against Japan—which the United States numbers among its possessions. Among them, the Marshall Islands are currently being used as a garbage dump.

China Tibet, an ancient colony of China, was annexed by China in 1958, forcing the tiny nation's spiritual leader, the Dalai Lama, into exile. Although Tibet's nationhood is unrecognized, Chinese troops enforce stiff martial rule, suppressing Tibetan nationalism with deadly force.

Russia Although there are still so-called autonomous nations inside Russia, the former Soviet Union has shed most of its former colonies, particularly the three Baltic states. Other questions of Russia's internal makeup are much in doubt, especially the region of Chechnya, which has been battling for independence since 1991.

Although the era of colonialism is largely over, these lands that remain in distant hands are a reminder that history is littered with land-grabbing takeovers. It is wise—and sobering—to remember that history is written, and maps redrawn, by the winners.

Where Is the "Third World"?

"Poor, strife-ridden, chaotic."

Russia and Yugoslavia during the past few years? Or the inner-city war zones of Los Angeles during the bloody riots of May 1992?

No. How about a definition of what it means to be part of the Third World. Those words were the standard set by Prime Minister Lee Kuan Yew of Singapore in 1969. Does that put Russia or Los Angeles in the Third World? That's not meant to be a funny question. But it is, sadly, a legitimate one.

"Third World" is a 1950s-vintage phrase invented by French intellectuals. The world needed a tidy reference point for the newly emerging independent nations in Asia and Africa. These former colonies, mostly poor and politically unstable, were christened *le tiers monde*. The First World was the West, dominated by the United States, Western Europe, and their satellites, most with free-market economies. The Second World was the Soviet Union and its East European allies, a world characterized by socialist systems with state-run economies. And the Third World tag was applied to the nonaligned and developing nations of Africa, Latin America, and Asia. In these developing nations at the time, average incomes were much lower than in industrial nations, their economies relied on a few export crops, and farming was at the subsistence level.

A glance at the following list of countries whose names changed in the years following independence is revealing. Most are part of what has been traditionally considered the Third World. Many of them also have the unhappy distinction of still being counted among the world's poorest nations by the United Nations. Of the forty-one nations singled out by the UN as the world's poorest, twenty-eight are in Africa, another eight are in Asia, four are Pacific Island groups and one is in the Caribbean.

But time and events have combined to make *Third World* a misused anachronism. From the very beginning, the concept of a monolithic Third World that was somehow separate and different from either East or West made little sense. In the first place, the notion of nonalignment was a fairly ridiculous one.

More significantly, the third-world label ignored vast differences of culture, religion, and ethnicity. The impoverished countries of Central America had little in common with sub-Saharan Africa. And what did equatorial Africa have to do with Afghanistan? To complicate matters, someone at the United Nations introduced the notion of a Fourth World, a term increasingly used to identify the poorest group of nations.

Huge changes in the world's geopolitical alignments have made these labels obsolete. In the first place, there is no longer a Second World in competition with a First World with which to contrast a "third." The fall of communism in the Soviet Union and Eastern Europe took care of that distinction.

Equally significant is the rapid economic expansion in many countries that once fit the description of a Third World country. Singapore of the 1990s is certainly no longer the "poor, strife-ridden, chaotic" island it was when the phrase was coined. South Korea, Taiwan, and Hong Kong all have robust economies. And in many other nations, there are growing chasms between rich and poor.

In that sense, the Third World did and does refer to something real, almost a state of being rather than a definable group of places. *Third World* still connotes vast social problems such as disease, hunger, and bad housing. But no group or country enjoys a monopoly on these dubious distinctions. The ills that have come to be associated with the Third World are present everywhere on the globe. These problems are usually at their worst in the large cities of Latin America, Asia, and Africa—from Mexico City and Rio de Janeiro to Cairo, Lagos (Nigeria), and Calcutta. News reports about conditions in the former Soviet

Union describe it as "third world," and Russia even seeks aid from South Korea. Elsewhere in Europe, Yugoslavia's ethnic fighting is as brutal as any "tribal" war in Africa. Even the United States is pocked with third-world enclaves, either in the inner cities or in rural hamlets, where Americans suffer life expectancies typical of Bangladesh, and infant mortality rates are worse than those found in many so-called Third World nations.

NAMES: Changing Names in the Twentieth Century

Before the European independence movements of the 1990s transformed the former Iron Curtain countries, an earlier wave of independence movements shook the world, principally during the 1960s.

In Africa, Asia, and South America, former colonies shook off the vestiges of European and American control. The result was an extraordinary number of place-name changes, frequently adding to the general confusion over matters geographic. As independence took hold, the people in these countries, like divorced women reclaiming their maiden names, shucked off their European-map references and returned to traditional names to reflect their newfound independence.

Countries that might have been familiar from the maps of the past disappeared in a flurry of rechristenings. The following is a list, by continent, of present-day countries and their dates of independence, along with their previous—and perhaps more familiar—colonial names. Those nations marked with an asterisk (*) are listed as among the world's poorest by the United Nations. (A guide to the freshly minted Europe can be found in Chapter 2, pages 94–98.)

Present Name	
(Date of Independence)	**Colonial Name**

Africa	
*Benin (1960)	Dahomey
*Botswana (1966)	Bechuanaland
*Burkina Faso (1960)	Upper Volta

*Burundi (1962)	German East Africa; Ruanda-Urundi
*Central African Republic (1960)	French Equatorial Africa
*Chad (1960)	French Equatorial Africa
Congo (1960)	French Equatorial Africa or Middle Congo
*Djibouti (1977)	French Somaliland
*Equatorial Guinea (1968)	Spanish Guinea
Ghana (1957)	Gold Coast
*Guinea (1958)	French West Africa
*Guinea-Bissau (1974)	Portuguese Guinea
*Lesotho (1966)	Basutoland
*Malawi (1964)	British Protectorate of Nyasaland
*Mali (1960)	French Sudan
Namibia (1990)	German South-West Africa
*Rwanda (1962)	German East Africa; Ruanda-Urundi
*Tanzania (1964)	Tanganyika and Zanzibar
Zaire (1960)	Belgian Congo
Zambia (1964)	Northern Rhodesia
Zimbabwe	Rhodesia

In 1965, Rhodesia declared itself an independent white-minority government free from Great Britain. Following a protracted guerrilla war, the white government agreed to democracy, leading to a black majority government in 1978. The name was later changed to Zimbabwe Rhodesia and independence was formally recognized in 1980. The country is now known as Zimbabwe.

Asia
*Bangladesh (1971)	East Pakistan
*Myanmar (1948)	Burma
Sri Lanka (1948)	Ceylon
Vietnam	French Indo-China

Vietnam was divided in 1954 into North Vietnam and South Vietnam. Following the fall of Saigon after the war in Vietnam, the country was unified under Communist rule in 1975.

Pacific Islands
*Kiribati (1975)	Gilbert Islands
*Tuvalu (1978)	Ellice Islands
*Vanuatu (1980)	New Hebrides

South America
Belize (1964)	British Honduras
Guyana (1970)	British Guiana
Suriname (1975)	Dutch Guiana

What Is the Fourth World?

Africa
  Cape Verde
  Comoros
  Ethiopia
  Gambia
  Mauritania
  Mozambique
  Niger
  Sierra Leone
  Somalia
  Sudan
  Togo
  Uganda
  Sâo Tomé and Príncipe

Asia
  Afghanistan

Bhutan
Laos
Nepal
Yemen
The Maldives

Caribbean
Haiti

Pacific
Western Samoa

IMAGINARY PLACES: Is There Really a Transylvania?

Geographic Voices   Bram Stoker, *Dracula* (1897)

I had visited the British Museum and made search among the
books and maps in the library regarding Transylvania: It had struck
me that some foreknowledge of the country could hardly fail to have
some importance in dealing with a nobleman of that country. I find
the district he named is in the extreme east of the country, just on
the borders of three states, Transylvania, Moldavia, and Bukovina,
in the midst of the Carpathian mountains; one of the wildest and
least known portions of Europe. I was not able to light on any map
or work giving the exact locality of the Castle Dracula, as there are
no maps of this country as yet to compare with our own Ordnance
Survey maps; but I found that Bistritz, the post town named by
Dracula, is a fairly well known place.

Few places in literature convey the instantaneous sense of dread,
mystery, and evil evoked by Bram Stoker's description of the foreboding
landscape of Transylvania. In Stoker's 1897 novel *Dracula*, the young
Jonathan Harker travels from England to this remote part of Europe,
the home of one of the most terrifying and fascinating characters in

fiction, Count Dracula. But does such a place as Transylvania actually exist, or is it merely the creation of Stoker's vividly eerie imagination?

In fact, Transylvania is on the maps, a historic region of central and northwestern Romania. It sits high on a plateau separated from the rest of Romania by the Transylvanian Alps to the south and the Carpathian Mountains to the east and north. The Carpathians form a natural boundary between Romania and Moldavia, one of the former Soviet republics, now independent.

Sitting along a natural invasion route for armies passing between Asia and Eastern Europe, Transylvania has had more than its share of bloody history. During its tempestuous past, Transylvania has been invaded, fought over, and controlled by everyone from the Huns to the Ottoman Turks, the Habsburgs, and the Austrian-Hungarian empire. During the Second World War, Romania aligned itself with Hitler's Germany. The Soviet Union invaded, and Romania, including Transylvania, became part of the Warsaw Treaty Organization, aligned with the Soviet bloc. Besides its unfortunate location as a convenient pathway for armies of conquest, Transylvania also boasts rich mineral deposits, large areas of forest, and fertile plains.

But if Transylvania is real, what about Dracula? The name, which means "demon," was first applied to Vlad IV (that stands for "the fourth" not "intravenous"), known as the Impaler. Vlad was a fifteenth-century local prince upon whom Stoker based the notorious character of Count Dracula. The atrocities committed by Vlad, who reigned in the region of Walachia from 1456 to 1462, are nearly unspeakable. In 1459, he attacked a small town and, for reasons unknown, systematically impaled thousands of locals on sharpened stakes. Vlad supposedly then ate a casual dinner in the midst of this grotesque scene of mass murder. Another of his legendary exploits was inviting beggars from the local countryside to attend a feast, locking them in a castle, and having it burnt to the ground. The number of Prince Vlad's victims is not known exactly, but it may have reached one hundred thousand before he was deposed and beheaded.

Vlad was not the only local prince to give Transylvania its blood-thirsty image. In 1514, an uprising of peasant farmers in the region was put down with brutal force. The leader of the revolt was then forced to sit on an iron throne that was heated and had a white-hot crown placed on his head. As this unfortunate rebel prince slowly roasted, his followers were forced to eat pieces of his cooked flesh. Transylvania's vampire image was further enhanced by the exploits of Countess Elizabeth Bathori. According to legend, she believed that bathing in the blood of virgins increased her beauty. Some six hundred fifty young peasant girls were reportedly killed for the sake of the noblewoman's vanity.

Transylvania, along with the rest of Romania, was ruled until recently by a Communist dictator who seems to have had a touch of the Transylvanian madman in his blood. Nicolae Ceausescu came to power in Romania in 1967 and began a reign of brutal suppression that fit right in with the region's gory past. He filled prisons with political prisoners who suffered modern forms of torture and death, and razed entire villages to force the people into dreary state-owned buildings while building a magnificent palace for himself. An army rebellion in 1989 led to Ceausescu's overthrow. He was quickly tried and executed, along with his wife, by firing squad. One tragically ironic footnote to this recent chapter of Romania's sad history is the awful result of a misguided experiment. In an attempt to improve health among Romanian children during the Ceausescu regime, infants were unknowingly given blood transfusions with AIDS-tainted blood, spawning a generation of Romanian AIDS babies.

How Did Japan Do It?

If you simply take the usual geographical factors into account, Japan should not be one of the world's most powerful nations, with an economy that ranks among the world's leaders. From a purely geographical standpoint, Japan has everything going against it. Yet in less than a century—after Commodore Matthew Perry sailed into Tokyo Bay for the second time in 1854—Japan transformed itself from an isolated, practically medieval feudal state into a modern, innovative, economic superpower.

Japan is small, made up of hundreds of mountainous islands scattered over more than 1,500 miles (2,400 km). It lacks natural resources—land and oil, most of all. Four fifths of the country is mountainous. Forests, which are held sacred, cover almost two thirds of the country, more than in any other industrial nation. Only 15 percent of Japan's total land can be farmed. Sitting on the Pacific Ocean's Ring of Fire, Japan is subject to regular violent earthquakes, volcanic eruptions, and devastating tsunamis. Japan also lies on the western Pacific typhoon track, which frequently brings fierce tropical hurricanes and deadly flooding to coastal areas.

But in a way that might be described as typically oriental, Japan has done a judo move on its geography, turning weaknesses into strength. All of the nation's geographical flaws have been met by positive responses. Rather than limiting the Japanese, geography has compelled the government and people to adopt cultural attitudes and national policies that have allowed the tiny island state to prosper. For instance, Japanese architects have designed skyscrapers for Japan's cities that can survive earthquakes of great strength. In cities situated near live volcanoes, elaborate evacuation drills are practiced regularly in which the Japanese genius—or obsession—for social organization and order is vividly demonstrated.

The fact that Japan is an island forced the Japanese to concentrate on developing trade. Today, its phenomenal wealth is primarily based on international trade. From a historical perspective, Japan is not unique in this respect. Other great empires throughout history have been centered on small islands or land-poor nations that turned to overseas trade to buttress their geographically limited home economies. In the process, they gathered technical, agricultural, artistic, and cultural expertise from the nations with which they had contact. This was certainly true of ancient Greece. While not entirely an island, Greece was made up of several Aegean island states which, along with the mainland city-states, had precious little land to exploit. Forced to turn seaward, the Greeks became great sailors and dominated Mediterranean trade. Then from every country they reached, the Greeks brought back substantial contributions to their mathematics, astronomy, philosophy, and art, all of which enriched their own civilization and were then passed on to the different sectors of their empire.

Similarly, Holland is not an island. But pressed up against a sea that constantly threatened to drown the countryside, and lacking sufficient land and resources, the Dutch still built a vast international sea-trading empire. Based in Amsterdam, the Dutch empire spread across the Pacific during its golden age in the sixteenth and seventeenth centuries. And finally Great Britain, a small island with few natural resources, rose to become one of history's greatest empires during the eighteenth and nineteenth centuries by dominating the seas and creating colonies that provided raw materials and markets for finished goods.

Each of these earlier great empires can be contrasted with the United States, which enjoyed great advantages in size and resources. But precisely because the country is so big, many American companies allowed the development of an aggressive world trade policy to take a backseat to the domestic market. Content to sell to a prosperous market at home, these American companies allowed other countries—Japan and Germany, the defeated nations of World War II, in particular—to develop overseas customers more effectively and efficiently. That "economic isolationism" has now caught up with the United States and the country is paying dearly—in such basic manufacturing industries as steel, automobiles, shipbuilding, and consumer goods—for its past failure to sell to the world.

In its post-1854 quest for modern power, Japan did not make the mistake of turning inward, as so many other countries have done, whether or not they were limited in space and resources. The leaders of Japan's government and industry spent the past century aggressively looking abroad. In the first forty years of the twentieth century, they did so violently, as the highly militarized, traditional Japanese government/business structure attempted to take by conquest what it needed—land and raw materials. In its war with Russia in 1904, Japan won control of the Korean Peninsula and gradually moved into Mongolia, Manchuria, and China during the 1930s. By 1940 it was ready to make a bolder move for control of the wealth and resources of the eastern Pacific area, a move that led to confrontation with American interests and the eventual attack on Pearl Harbor. But defeat in World War II left Japan a ruined, occupied nation in 1945. And the devastating losses of people, property, and overseas possessions presumably should have doomed Japan's goal of global superpower status. So what explains the Japanese economic miracle?

First of all, Japan could more or less feed itself, in spite of the shortage of arable land. That shortage forced the Japanese to turn to a self-sustainable diet. Raising cattle for beef or dairy is space-intensive because cattle demand lots of room to graze. They also eat tremendous quantities of grain and consume large amounts of water. In Japan, the pressures of space mean few cattle are kept—the Japanese eat no milk-based products and the comparatively little beef they consume is imported. Rice, soy, and fish have always been the nation's traditional staples. Fish are "farmed" locally in ponds, caught in coastal waters, or brought in by Japan's vast international fishing fleets. In the past, Japan has been self-sufficient in growing rice. The warm, wet summer climate in the South is ideal for rice. The Japanese government has also taken a strong hand by heavily subsidizing rice farming. Overseas rice growers, like those in the United States, have learned how heavy that helping hand is.

Another factor in Japan's rise is the country's cultural attitudes, which geography has played a critical role in shaping. The traditional Japanese notion of *kazoku* (literally, "family" or "harmony") was cultivated over thousands of years, uninfected by foreign concepts, including those of Western religions or the European Enlightenment that place so much more emphasis on individual freedoms. It is this Japanese tradition of fidelity and loyalty to authority that is at the heart of the relationship between Japanese firms and their employees, bolstering productivity and promoting highly cooperative labor relations—rather than the worker-management antagonism typical of Western labor relations.

Ironically, one of the consequences of America's wartime victory over Japan created enormous dividends for the Japanese. Written during the American postwar occupation, the Japanese constitution stipulated that Japan could not spend more than 1 percent of its gross national product on defense and the military. While everyone else in the world, led by the United States and the Soviet Union, devoted huge portions of their national budgets to the massive Cold War arms race and involvement in costly local wars from Vietnam to Afghanistan, the Japanese were investing in business and technology what they could not spend on tanks and aircraft carriers.

These investments helped fuel Japan's astonishing rise, which had three distinctive phases. First came the development of heavy industry, which soon surpassed Japan's traditional textile production in export earnings. By 1956, Japan had surpassed Great Britain as the world's major shipbuilder. The second phase was the boom in consumer electronics and automobiles. Again, geography played a role. With limited resources and little space,

Japanese manufacturers concentrated on building everything smaller—which explains their domination in the field of miniaturized transistors. Completely dependent on overseas oil, Japanese automakers led the way in developing fuel-efficient cars. American automakers kept blithely turning out the profitable gas guzzlers that had been popular when oil production was not in the hands of OPEC, a cartel that could turn off the tap at will. By the mid-1970s, Japan had captured 21 percent of the world's automobile production, second only to America's Big Three. The third, most recent stage is characterized by knowledge-intensive products, with massive research and development going into computers and biotechnology. The Japanese have coupled high-technology products with the application of computer-controlled manufacturing techniques to achieve enormous productivity gains and product dependability.

At the same time Japan was building these areas at home, it was channeling its overseas investments into the low-wage economies of other Asian nations. This strategy not only lowered production costs—keeping prices low and profits high—but it created overseas markets in which Japan's products could be sold.

The Japanese have also led the way in proving that economic development is far more significant to political stability than force of arms. Providing these Southeast Asian nations—many of them pawns in the superpower confrontations of the sixties and seventies—with jobs and economic security has politically stabilized the entire region. Even Vietnam, one of the most tragic symbols of the cold war's futility, has quietly become a consumer society in which the former confrontation between America and the Communist bloc has given way to more basic desires for a higher standard of living.

Geographic Voices   Marcel Junod visiting Hiroshima, September 9, 1945, from *Warriors Without Weapons* (1951)

At three-quarters of a mile from the centre of the explosion nothing at all was left. Everything had disappeared. It was a stony waste littered with debris and twisted girders. The incandescent breath of the fire had swept away every obstacle and all that remained upright were one or two fragments of stone walls and a few stoves which had remained incongruously on their base.

We got out of the car and made our way slowly through the ruins into the centre of the dead city. Absolute silence reigned in the whole necropolis.

Geographic Voices   Simon Winchester, *The Man Who Loved China*

Chongqing—or Chungking as it was written—is now like few places on earth, growing so fast and furiously that it is hard to keep up with the speed of the changes. Chongqing is now the most populous city in China. . . . Thirty-eight million people live crammed within its metropolitan limits. The frantic rhythms of their lives capture the concentrated essence of everything, good and ill, about the awe-inspiring, terrifying entity that is today's new China. . . . Now eight bridges sweep over the [Yangzi] river, and eight sparkling new monorail lines on stilts run along beside it. Clusters of skyscrapers have sprung up in each of the half dozen commercial centers . . . There are teeming masses of people, happy-looking, prosperous, loud, boisterous, well-dressed, well-coiffed, and well-fed, and all Chinese. . . . All seems happy. All are watched.

Milestones in Geography VI
1950 to the Present

1950 The world population is approximately 2.3 billion people.

1951 The Korean War begins as the Army of Communist North Korea invades South Korea. Fighting continues until 1953; a treaty leaves the Korean Peninsula divided between North Korea and South Korea.

1953 Sir Edmund Hillary of New Zealand and Tenzing Norkey of Nepal reach the summit of Mount Everest.

1953 Egypt becomes an independent republic under military rule.

1955 The Federal Republic of Germany (West Germany) becomes a sovereign state.

1955 The Warsaw Pact is established, linking the Soviet Union and the nations of Eastern Europe in a military alliance against the West.

1956 The Sudan gains its independence.

1959 British anthropologist Louis Leakey (1903–72) finds fossil remains of an early hominid from about 1.75 million years ago while working in Olduvai Gorge, Kenya. He names it *Zinjanthropus*, but later it is called *Australopithecus robustus* or *A. boisei*.

1959 Following a revolution, Cuba becomes a Marxist state under the leadership of Fidel Castro.

1960 The Belgian Congo gains its independence. (Later renamed Zaire.)

1960 Cyprus becomes an independent republic.

1961 Louis Leakey and his wife, Mary Leakey, find the first fossil remains of *Homo habilis* or "handy man."

1961 The Berlin Wall is constructed, dividing Communist East Berlin from the rest of Berlin, a city that is inside Communist East Germany.

1962 Uganda and Tanganyika gain independence.

1962 France grants independence to Algeria.

1964 The Aswan Dam on the Nile is completed. It creates the vast Lake Nasser, and is used for irrigation and production of hydroelectric power, meeting half of Egypt's electricity needs. However, it brings profound ecological changes as well. Floodwaters no longer fertilize the land, which increases the demand for chemical fertilizers.

1964 Kenya becomes a republic; Tanganyika and Zanzibar unite to form Tanzania; Northern Rhodesia becomes the independent republic of Zambia.

1965 Gambia gains independence.

1965 Under a white minority government, Rhodesia declares its independence from the United Kingdom.

1966 Colonial British Guiana, located on the north coast of South America, becomes the independent republic of Guyana.

1967 In the Arab-Israeli Six-Day War, Israel gains control of the Sinai Peninsula from Egypt, the Golan Heights from Syria, and the Jordan River's West Bank; all of Jerusalem, once partly controlled by Jordan, is united under Israeli rule.

1968 Soviets invade Czechoslovakia to crush its liberal government.

1970 Celebration of the first "Earth Day," aimed at increasing awareness of environmental dangers.

1970 After a long, unsuccessful war for independence, the region of Biafra surrenders to Nigeria after millions have been starved by the civil war.

1971 Bangladesh, formerly East Pakistan, becomes independent.

1971 The United Nations seats Communist China and expels Nationalist China (Taiwan).

1973 The Yom Kippur War. Arab armies strike Israel during the highest holy day. Egypt retakes a portion of the Sinai Peninsula. An Israeli counterattack into Egypt is halted by a ceasefire.

1974 A team led by Donald Johanson and Maurice Taieb discover 40 percent of the skeletal remains of an early hominid that is more than three million years old in the Afar region of Ethiopia. Named Lucy because the Beatles' song "Lucy in the Sky with Diamonds" was playing at the time of the discovery, the remains are representative of a previously undiscovered species that Johanson names *Australopithecus afarensis*.

1975 The world's population reaches an estimated 4 billion people.

1976 North and South Vietnam are reunited under Communist control after twenty-two years of separation. Hanoi is the new capital and Saigon is renamed Ho Chi Minh City.

1976 The Central African Republic, a former French colony, is renamed the Central African Empire by Emperor Bokassa I, who installs himself in a lavish $25 million ceremony. Bokassa is later deposed and the country reverts to the Central African Republic.

1979 The Camp David peace accord negotiated by President Jimmy Carter brings an end to the thirty-year war between Egypt and Israel, marking the first Arab-Israeli peace treaty. Under its terms, the Sinai Peninsula, occupied by Israel since 1967, is returned to Egypt.

1979 Iran becomes an Islamic republic led by Ayatollah Khomeini following the deposing of the U.S.-backed shah of Iran.

1979 The leftist-rebel Sandanistas take control of the Central American nation of Nicaragua. Elected in 1980, Ronald Reagan commits his administration to the overthrow of the Sandanistas by supporting rebel fighters known as contras.

1980 Iraq attacks Iran, beginning a bloody ten-year war that ends inconclusively.

1982 The Falklands War. Argentine forces invade the small islands off its coast to which it claims ownership, but are defeated by British troops.

1983 In Kenya, Dr. Meave G. Leakey finds a fossil jaw tentatively dated at sixteen to eighteen million years ago, and identified as *Sivapithecus.*

1984 The first hole in the ozone layer of the atmosphere is observed over Antarctica.

1984 An expedition in Kenya led by Andrew Hill finds a jawbone from an *Australopithecus afarensis* that is dated to five million years, the oldest known representative of the hominid line.

1984 Working in the Kola Hole in Siberia, Soviet researchers drill the world's deepest hole, going to a depth of 7.5 miles (12 km) and reaching the earth's lower crust.

1986 President Ferdinand Marcos flees the Philippines after ruling for twenty years; he is succeeded by newly elected president Corazon Aquino.

1986 Paleontologists Tim White and Donald Johanson locate 302 pieces of a female *Homo habilis,* known as OH62, now known to be 1.8 million years old.

1987 The Airborne Antarctic Ozone Experiment suggests that chlorofluorocarbons (CFCs) are responsible for the ozone hole, leading to calls for international action against CFCs.

1988 French and Israeli scientists discover fossils in a cave in Israel that are the remains of a 92,000-year-old modern-type *Homo sapiens*, more than doubling the length of time that modern humans are thought to have existed.

1989 Tens of thousands of pro-democracy students protest against harsh authoritarian rule in China's Tiananmen Square until Chinese authorities brutally repress the dissent.

1989 The Berlin Wall is opened after twenty-eight years, allowing free movement between the two halves of the formerly divided city.

1989 In Czechoslovakia, the parliament ends the Communist Party's dominant role.

1990 The Yugoslavian Communist Party ends its monopoly on power.

1990 Nelson Mandela is released from a South African prison; South African president F. W. de Klerk calls for an end to the apartheid system.

1990 Iraqi troops invade Kuwait; the United States dispatches troops to defend Saudi Arabia from attack.

1990 East and West Germany are reunited.

1990 Labor leader Lech Walesa is elected president of Poland.

1991 The United States and its allies attack Iraq; Iraqi forces are expelled from Kuwait.

1991 The Warsaw Pact ends, dissolving the military alliance among the Eastern European nations.

1991 The three Baltic republics of Estonia, Lithuania, and Latvia win their independence from the Soviet Union.

1991 The remaining twelve Soviet Republics all declare their independence, bringing an end to the USSR.

1992 A *Homo erectus* jawbone found in the former Soviet Republic of Georgia is tentatively dated to 1.6 million years. Its existence suggests that human ancestors may have left Africa several hundred thousand years earlier than previously thought. If the early date is confirmed, this would be the first solid evidence that human ancestors spread out of Africa earlier than one million years ago.

1992 Riddled by civil war after the demise of Communist rule, Yugoslavia is split up into the new republics of Slovenia, Croatia, Bosnia and Herzegovina, and Macedonia. The regions of Serbia and Montenegro retain the name of Yugoslavia.

1992 A skull fragment found twenty-five years earlier near Lake Baringo, Kenya, is dated by Dr. Andrew Hill to 2.4 million years. This date extends by half a million years the age of the genus that led to and includes modern humans.

What Can You Build with BRICS?

When leaders of five recently emerging world economic powers convened in New Delhi in March 2012, the focus was mostly business—mutual trade and reforms of the global financial system as the global economy suffered through another year of fits and starts. But they also called for diplomacy in addressing the violence in Syria and Iran's disputed nuclear ambitions.

This is all very remarkable considering the histories and status of these five countries—Brazil, Russia, India, China, and South Africa, hence the acronym BRICS—in fairly recent history. It was not so long ago that Brazil was a sputtering "third world" economy ruled largely by military dictators until democracy returned in 1989.

Russia—technically the Russian Federation—emerged from the 1991 breakup of the Soviet Union, the bitter Cold War antagonist of the United States and Western Europe. Adopting a new constitution in 1993, Russia was still beset by widespread corruption, political turmoil from independence movements and stark economic difficulties largely the result of little investment under decades of Communist rule that saw a large portion of its economy poured into its military. But the nation once feared for the tanks and armies displayed in Moscow's May Day parades or rolling through captive Eastern bloc nations, and for the missiles it once tried to place in Cuba and aimed at the West, Russia was now being watched for the power it wielded through control of enormous natural resources and a new generation of Russian capitalists ready to assert the role of the world's largest nation in the new global economy.

India, among the most ancient of the world's civilizations, had long been the poster child of starving, undeveloped countries. But fueled by technology and a growing middle class that prized education, India, with its billion-plus population, was transformed into an economic giant at the beginning of the twenty-first century.

China, in 2010, overtook Japan as the world's second-largest economy—a stunning development for a nation that was poor and backward for much of the last half of the twentieth century. By 2010, half of the world's ten largest banks were Chinese. (What Chairman Mao, China's authoritarian Marxist leader, would make of this development is an extraordinary thought to ponder.) China also surpassed the United States as the world's top energy consumer. Chinese factories were humming and newly minted Chinese consumers wanted to buy their own cars. Lots of them. In *The World Is Flat*, Thomas Friedman reports that during a visit to Beijing in 2004, he learned that the people of that city alone were buying thirty thousand cars *a month*. China is now also one of the world leaders in the dubious category of creating traffic jams and air pollution.

The BRICS group grew out of the earlier gathering of Brazil, Russia, India, and China, the so-called BRIC nations, a phrase coined by a prominent investment banker in a 2001 paper that forecast the growing power of these four nations. In 2010, South Africa, once an international pariah because of its apartheid policies, joined the group. Condemned by much of the world for its racist political and cultural policies, South Africa was transformed in 1990 when the apartheid laws were dismantled. In 1994, Nelson Mandela, freed earlier from a South African jail, became the country's first black president.

The 2012 New Delhi meeting of the BRICS nations was a flexing of their growing economic muscle. Representing nearly three billion people, these five countries have enjoyed generally fast-growing economies, and aspire to have more say in the global financial system long dominated by the United States and Europe. Part of that effort involves their desire to reform the International Monetary Fund and the process for selecting the World Bank's president. Which brings us to the next question . . .

Does the World Bank Have ATMs?

In the waning days of World War II, the forty-four Allied nations sent delegates to a financial conference at the Mount Washington Hotel in Bretton Woods, New Hampshire. The goal of the more than seven hundred delegates to the Bretton Woods Conference, as it was called, was as lofty as nearby Mt. Washington, highest peak in the Northeast United States (6,288 feet; 1,917 meters). They were there to attempt to stabilize the world's financial and currency markets after the devastating war.

Two key creations of the conference remain in the headlines today— the World Bank and the International Monetary Fund, or IMF. (But the quick answer is, "No, don't look for a World Bank ATM on the corner.")

The World Bank is now made up of 187 member countries whose mission is to help poor nations develop their economies and improve social conditions. Founded in 1944, it is headquartered in Washington and has more than one hundred offices worldwide. Since its inception, World Bank tradition holds that the United State selects the organization's president and Europe chooses the head of the IMF. But developing nations and emerging powers have been pushing for a greater role.

The World Bank has long been criticized because the so-called free-market reform policies it typically advocates—which include deregulation, privatization, and downscaling of government—can often hurt economic development if they are implemented poorly or too quickly. The other chief criticism of the World Bank is its tradition of American leadership. While the World Bank represents 186 countries, it has been dominated by a small number of countries, especially America and the European powers. The increasing resistance to that tradition is a sign of the growing economic power of the some of the formerly developing nations that now want a place at the World Bank's boardroom table.

Created at the same conference, the International Monetary Fund was intended to prevent a reoccurrence of the Great Depression, when countries raised trade barriers in an attempt to protect national economies. Doing so added to a great inflation problem and chaos in currency exchange that brought the international banking system to its knees, worsening the worldwide depression of the 1930s.

After the war, the allied nations were primarily concerned with rebuilding Europe—war-torn France received the first loan—and restoring the economic system. Among the delegates to the conference was the British economist John Maynard Keynes, who wanted the IMF to act as a fund that would help create economic activity; others wanted the IMF to function as a more traditional bank that made loans that had to be repaid. It is an old argument that was being replayed in 2012 in both Europe and the United States, as proponents of economic stimulus argued with those who want more austerity as a way of forcing governments to live within their means.

The IMF was credited with helping to rebuild the system of capitalism after the war, but still allowing for human welfare. Today, the number of IMF member countries has more than quadrupled—from the 44 states involved in its establishment to 188 today—as former colonial African countries and the Eastern European nations once controlled by the Soviet Union have been invited to join.

Like the World Bank, the IMF has been criticized by some economists for demanding too much austerity from borrower nations. This controversy was at the heart of the political chaos in Greece in 2012 when the Greeks began to balk at the extreme austerity programs forced on the country by its lenders, leading to the fall of the government.

Both the World Bank and the International Monetary Fund, although much changed since 1944, reflect the postwar global economy that began to grow and has only accelerated in the early years of the twenty-first century. The growing controversy over who holds the levers of power at these institutions reflects how the world order has changed remarkably in the past few decades.

What Was the Arab Spring?

A wise man once sang, "You don't need a weatherman to know which way the wind blows." Bob Dylan's "Subterranean Homesick Blues" spoke to a generation of American protesters in the 1960s. But those words applied equally to a howling wind of protest that was a far cry from any weatherman's ideal springtime.

While the changes affecting the BRICS nations and the emergence of China as an economic powerhouse were breathtaking in scope and suddenness, perhaps the greatest upheaval of the world order in the early days of the twenty-first century took place in some of the oldest nations in civilization.

Beginning in late 2010, a wind of unexpected power, protest, resistance, and rebellion blew across North Africa and through other Middle Eastern Arab states with something like hurricane force. The winds swept away some of the most entrenched dictatorships that had ruled for decades in such countries as Tunisia, Yemen, Libya—where NATO forces helped topple strongman Muammar al-Qaddafi, captured and killed by rebels—and Egypt, the most populous Arab state, where military dictator Hosni Mubarak had ruled for thirty years. He was deposed, tried, and sentenced to life in prison in 2012.

The successive waves of demonstrations and protests that came to be known as the Arab Spring (or the Arab Awakening or Arab Uprisings) actually began in December 2010 and continued to roil through the Arab world well into 2012 in such countries as Syria, where massacres of civilians continued to be reported in June of that year.

The protests, strikes, rallies, and demonstrations had affected many other countries in the area, including Lebanon, Saudi Arabia, Mauritania, and Oman, among others. In most cases, the protests were met with violent opposition by military authorities and pro-government militias.

In each of these Arabic countries, new social media sites spawned by the Internet age such as Twitter and Facebook, and the widespread availability of cell phone technology, clearly helped spread the groundswell of protests and were used as tools to organize demonstrations and even military operations. Television stations, including the Al Jazeera Arabic-language news network, were also instrumental in fostering the wave of revolution around the region.

But more fundamentally, what these extraordinary and historic protests shared was a lingering frustration over stagnant economies, limited economic opportunities, corrupt and often brutally repressive regimes, and a fundamental lack of democracy along with basic civil and human rights.

Although the long-range impact of the Arab Spring is impossible to gauge, an entire region that had been focused on the Arab-Israeli conflict for decades, was instead undergoing revolutionary upheaval. It was in one sense the triumph of technology and the force of globalization that shaped the Arab Spring in a revolution whose speed was almost breathtaking but which was far from over.

As Thomas Friedman of the *New York Times* wrote in 2011, while the Arab Spring reached a climax, "Surely one of the iconic images of this time is the picture of Egypt's President Hosni Mubarak—for three decades a modern pharaoh—being hauled into court, held in a cage with his two sons and tried for attempting to crush his people's peaceful demonstrations. Every leader and C.E.O. should reflect on that photo. 'The power pyramid is being turned upside down,' said Yaron Ezrahi, an Israeli political theorist."

Geographic Voices   In *Why Nations Fail*, a book that examines the huge disparities in incomes and living standards around the world, Daron Acemoglu and James Robinson commented on the Arab Spring, specifically in Egypt.

The roots of discontent in these countries lie in their poverty. The average Egyptian has an income level of around 12 percent of the average citizen of the United States and can expect to live ten fewer years: 20 percent of the population is in dire poverty. Though these differences are significant, they are actually quite small compared with those between the United States and the poorest countries in the world, such as North Korea, Sierra Leone, and Zimbabwe, where well over half the population lives in poverty. . . . Why is Egypt so much poorer than the United States? What are the constraints that keep Egyptians from becoming more prosperous . . . To Egyptians, the things that have held them back include an inefficient and corrupt state and a society where they cannot use their talent, ambition, ingenuity and what education they can get. But they also recognize that the roots of these problems are political. All the economic impediments they face stem from the way political power in Egypt is exercised. . . . This . . . is the first thing that has to change.

5

Paradise Lost? Geography, Weather, and the Environment

Viewed from the distance of the moon, the astonishing thing about the Earth, catching the breath, is that it is alive. The photographs show the dry, pounded surface of the moon in the foreground, dead as an old bone. Aloft, floating free beneath the moist, gleaming membrane of bright blue sky, is the rising Earth, the only exuberant thing in this part of the cosmos. If you could look long enough, you would see the swirling of the great rifts of white cloud, covering and uncovering the half-hidden masses of land. If you had been looking a long, geologic time, you could have seen the continents themselves in motion, drifting apart on their crustal plates, held aloft by the fire beneath. It has the organized, self-contented look of a live creature, full of information, marvelously skilled in handling the sun.

—Lewis Thomas
The Lives of a Cell

Who, What, or Where Is Gaia?

*If Physicists Study Physics, Do Meteorologists
    Study Meteors?*

If This Is Aphelion, Why Is It So Hot?

What's So Hot About the Equator?

"Everybody talks about the weather, but nobody does anything about it."

When Charles Dudley Warner (1829–1900) wrote these words in an 1897 *Hartford Courant* editorial, he obviously did not realize that people have been doing something about the weather for a very long time. Alone among the millions of species that have occupied the earth, *Homo sapiens* has been the only one with the knack to shape the earth and alter its climate on a grand scale. Like volcanoes, earthquakes, or a deadly asteroid from space, people have been a natural catastrophe.

This chapter examines the close relationship between geography, weather, and the environment. But it also looks at the impact that humanity is increasingly exerting on both the weather and the environment.

Since the time irrigation ditches and canals were first dug to control water and extend nature's capacity to make the land fertile, to the massive destruction of the world's rain forests in our lifetime, people have attempted to control and alter nature, always with some measurable impact on the earth—sometimes for good and sometimes with terrible results. But this simple fact remains: we are the only species that holds in our hands not only the destiny of other species but our own future as well. In a very short space of geological time, people have played havoc with the earth's natural resources. And we are very nearly at a point of no return.

Conservationism or environmental concerns didn't start with the first Earth Day in 1970. The first generation of people with a passionate concern for the preservation of the world's resources and wilderness areas was already active in the late nineteenth century. Both the Sierra Club and the National Geographic Society are more than a hundred years old. Acid rain was first described in 1872, and the greenhouse effect was identified in 1896. But for the past forty years, scientists and concerned environmentalists have stepped up their warnings about the possibly irreparable harm being done to the earth and its sensitive systems, all of which are required to sustain life. Unfortunately, these warnings have largely gone unheeded, outweighed by short-term political considerations, the profit motive, and superpower rivalries.

Only slowly has the world come to recognize the necessity to change the ways in which we utilize the earth's resources. In June 1992, many nations sat down at the Rio Conference on the environment in Brazil. Their ambitious hopes of finding a common solution to protect the earth's resources while fostering worldwide economic development were complicated by powerful national and regional interests, each with an agenda. No country approaches the United States when it comes to emissions of carbon dioxide and other threatening gases. America's per-capita consumption of energy and its carbon emissions are far higher than those of any other country. But the Chinese may eventually overtake the United States as the largest culprit in the emission of greenhouse gases, a result of the massive industrial use of China's vast coal reserves. Pressed to restrain their coal burning, the Chinese argue that coal is the only choice they have if they are to become a modern industrial nation. With more than a billion mouths to feed, the Chinese authorities say that their national economic need for continued coal burning outweighs concerns about future environmental damage. They argue that raising the standard of living must come before cleaning the environment.

But China is not a lone environmental outlaw. They have plenty of company. Any serious attempts to come to grips with such issues as the depletion of the ozone layer, the disappearance of rain forests, the real likelihood of global warming, and the international calamity of acid rain, are confronted by a maze of special interests and powerful countries willing to dismiss long-term environmental damage when faced with short-term constraints on their economies. In almost every country in the world, powerful forces have the ear of government leaders. These are people who often don't want to look past next year's shareholder reports. They have successfully painted even the most serious and conservative environmentalists as "save the whales" loonies who are more concerned over a few tiny snail darters or spotted owls than the jobs of thousands of workers. These are the same special interests and politicians who have stunted research and development of renewable energy sources and improved, conservation-minded technologies. The simple truth is that developing these new fields and technologies can create thousands of new jobs in high-growth, high-tech industries instead of preserving a relatively small number of old jobs in dying businesses.

Twenty years ago, when faced with the first OPEC oil crisis, the United States and the West had an opportunity. The industrialized nations could have committed to the development of massive new alternative-fuels programs, similar in scope to the effort that put men on the moon in the space of a decade. While some halting movements toward conservation were initiated and some countries have undertaken ambitious energy-conservation programs, a grand design fell by the wayside, the victim of shortsighted reliance on quick fixes to get the Arabs to lower the price of oil.

When the Iranian political crisis of 1978 sent oil prices skyward once more, America's president Carter started the United States on a more ambitious synthetic-fuel and solar-energy development program. But subsequent Republican administrations gutted those programs, expecting private industry to take the lead, and turning to "market forces" to initiate research and development. Needless to say, twenty years of opportunity have largely been squandered. American automakers have been dragged kicking and screaming through every attempt to improve gasoline efficiency or to switch to alternative-energy autos. Efficient, high-speed mass transportation has been put at the bottom of the list of the government's priorities. In 1992, Americans are still desperately addicted to foreign energy sources that fuel the greenhouse effect and leave the American economy exposed to another regional upheaval such as the Iraqi invasion of Kuwait.

Then there are people all over the world at the other end of the spectrum, those who are barely subsisting. There is a real difference between the rich industrial nations trying to sustain their comfort level and poor nations struggling to survive. For families in the poorest countries, the issue is simply to get through the next day. They don't have time to worry about the ozone hole and the end of the rain forests or the fact that the cutting of trees for firewood is hastening the spread of deserts and increasing the likelihood of deadly floods.

Science has increasingly come to understand that the earth is a tightly interconnected collection of organisms. For every link that is removed or altered, the chain of life gets a little weaker. If we are to make it as a species without taking too much of creation down with us, we had better start understanding these connections and working on making them stronger right now.

Who, What, or Where Is Gaia?

Almost every culture has held a view of the earth as a living, all but sentient thing, not simply a place where life happened to spring up. For want of a better term, we've called it Mother Nature.

In 1972, James Lovelock, a British scientist who had worked for NASA on moon and Mars projects, gave our living organism Earth a different name: Gaia, the name of the Greek goddess of Earth. In his books *Gaia: A New Look at Life on Earth* and *The Ages of Gaia: A Biography of the Living Earth*, Lovelock promoted his hypothesis that the earth is an immense living organism, not simply a big hunk of rock surrounded by gases. Lovelock's Gaia is self-regulating and self-changing. As Lovelock writes in *The Ages of Gaia*, "Gaia theory forces a planetary perspective. It is the health of the planet that matters, not that of some individual species of organisms. . . . The health of the Earth is most threatened by major changes in natural ecosystems."

Lovelock's Gaia hypothesis is not widely accepted by mainstream scientists. But name a new theory that the mainstream ever accepted straightaway. The view that the earth is, in Lovelock's words, "a complex system which can be seen as a single organism and which has the capacity to keep our planet a fit place for life" can be viewed as fringe science, a daring new vision of the relationship of all living things, or simply as an appropriate metaphor for the earth and its complex systems. But Gaia's ability to "heal" itself is being challenged. And one of Lovelock's major conclusions seems inescapable: "We need a general practitioner of planetary medicine. Is there a doctor out there?"

If Physicists Study Physics, Do Meteorologists Study Meteors?

Meteors are fragments of solid matter that enter the upper atmosphere and become visible as they burn up from the friction of air resistance.

When seen in the night sky, they are called shooting stars. Most of these meteors are smaller than a grain of sand and hit our atmosphere every day. But only a very few reach the ground. When they do, they are called *meteorites*. And when they do make it all the way down to earth, meteorites strike at tremendous speeds—as much as 90,000 miles per second.

So what do rocks falling out of the sky have to do with the weather? After all, *meteorology* is the science that investigates the weather and climate in the earth's atmosphere. (Weather is the day-to-day atmospheric conditions for a certain time and place; climate is day-to-day weather conditions over an extended period of time.) The guys who keep track of meteors are usually astronomers. The connection between *meteor* and *meteorology* is in the Greek origin of both words. *Meteor* comes from the Greek meaning "astronomical phenomenon" or "high in the air." *Meteorology*, on the other hand, is the "discussion or study of an astronomical phenomenon," specifically the phenomenon we call weather.

Weather is like a stew hung over a cooking fire. Change the ingredients even slightly or adjust the cooking temperature and you get a very different kettle of stew. The earth's weather-stew has five major basic ingredients: the sun, the earth's shape and position in space, its rotations, the atmosphere, and the oceans. Then local geographic conditions provide the seasonings that give this weather-stew its regional flavor.

It all starts with the sun, which provides the heat for cooking. Solar heat is not only the fire under the stew pot but also the spoon that stirs the kettle. Sunlight reaching the earth is either absorbed by the blanket of the atmosphere or reflected back into space. Geographic factors have much to do with the amount of heat that is absorbed. Snow and ice absorb very little heat, reflecting 75 percent of sunlight back into space, one of the reasons the polar regions are so cold. Dry sand absorbs 75 percent of this solar heat. That's why you can't walk barefoot on a sandy desert or the beach on a hot, sunny day.

The air in earth's atmosphere is heated by contact with the warm earth. As the air warms, it rises like a hot-air balloon and is replaced by cooler air that flows underneath. This infusion of cold air is also heated and begins to rise, beginning an endless circular motion of the air called *convection*. This cycle sets the air in motion, causing local winds and breezes. Of course, the air is heated differently according to local surface conditions—ocean, land, and altitude all affect the degree of heating—which is another of the ways geography affects the weather.

If the earth didn't spin, the winds would stay in neat little circles, all moving back toward the equator with equal speed. But the energy of the earth's rotation, called the *Coriolis force*, bends the winds and gives them their characteristic patterns. Winds at different latitudes follow predictable and regular movements in bands that are similar in both hemispheres. The *trade winds* blow steadily toward the equator in two zones, north and south of the equator. European sailors could rely on the "northeast trades" to provide a predictable wind for reaching the Americas. Where these two bands meet over the equator, the air is heated and rises, creating an area of calm known by sailors as the *doldrums*. Another typically calm, windless band occurs near latitude 30°, which sailors christened the *horse latitudes* because ships carrying horses from Spain were often becalmed for weeks. As animal feed ran out, the animals died and were thrown overboard, littering the sea with the carcasses of horses. After the horse latitudes, there is another reliable band of winds, the *prevailing westerlies*, which basically flow from west to east around the world between 30° and 60° latitude, both north and south of the equator. They are not as constant and strong as the trades, but are comparable to a meandering river of air. The third broad pattern, again repeated in both north and south, is the *polar easterlies*, cold, dry winds that move from east to west above latitude 60°.

These winds, the earth's rotation, and the placement of the continents also cause the great ocean currents. These currents move huge masses of water—sometimes cold, sometimes warm—from region to region. As they move, ocean currents have a great influence on climate. That's why the United Kingdom, which is warmed by the North Atlantic drift originating off Florida, has a milder climate than Labrador, which is on the same latitude but is cooled by a current from the Arctic Ocean.

Another factor of local geography with great impact on climate is the combination of mountains near the coast. Warm, moist winds rise on the windward side of mountains. As they rise, they expand and cool, causing the water vapor they carry to condense and fall as rain. The dry, cool air descends on the other side of the mountains. As it does so, it is compressed and warmed. The resultant dry, warm wind creates a markedly different climate and vegetation, as in California's Sierra Nevada or in Tibet, a dry region only hundreds of miles from a very rainy area. On one side of the mountains, the coastal side, rain is usually plentiful. But on the other side, the land turns arid.

If This Is Aphelion, Why Is It So Hot?

If the earth were truly a sphere—it isn't—and it was smooth—it isn't—and it wasn't tilted to one side—it is—it would be a much different place in which to live. But all of these peculiarities of the earth contribute to its uniqueness as a place able to sustain life. Change any one of these factors even slightly and you would have a very different earth.

Of course, the greatest factor, not only in our weather but in our very existence, is the sun, the explosive ball of gas located some 93 million miles distant. But that's another of the earth's quirks. We aren't always 93 million miles away from the sun. As the earth makes its annual orbit around the sun, our path is not a perfect circle. It is actually an ellipse. Sometimes we are closer, sometimes farther away from the sun. The seemingly obvious conclusion to draw from this quirk is that it is summer when we are closer to the sun and winter when we are farther away. That seems logical; but it's exactly opposite the truth. At *aphelion*, the word for the period of a planetary orbit farthest from the sun, it is summertime, at least in the northern hemisphere. (*Aphelion* comes from the Greek words *apo*, "away from," and *helios*, "sun.")

This seemingly illogical fact is due to the tilt of the earth's axis away from a parallel with the sun. The 23.5-degree tilt is the factor that produces our seasons. At all times, the sun's rays hit one half of the earth. But in the course of a year, when the Northern Hemisphere is tilted toward the sun, the sun's rays hit this half of the earth more directly. The earth is actually farther away from the sun during this period, which lasts from the vernal equinox in March—when the sun is directly overhead at the equator, giving the earth equal amounts of day and night—through the "summer" months of June, July, and August until September's autumnal equinox—when the sun is again directly overhead at the equator.

After the autumnal equinox, the earth has turned in its yearly orbit and the Northern Hemisphere begins its tilt away from the sun, receiving fewer direct solar rays. This creates the shorter days of the Northern Hemisphere's winter—but now it is summer in the Southern Hemisphere. Because of this tilt away from the sun, the North Pole or Arctic is plunged into near darkness while Antarctica enjoys near total daylight—or the midnight sun. Of course, the situation then reverses itself, season after season, year in year out.

What's So Hot About the Equator?

This is fairly simple. The equator is the one place on earth where the sun's rays are unaffected by the tilt of the earth. The equator is always exposed to the sun and receives direct solar rays year-round in spite of the season.

This means air at the equator is warmed more than at the poles. The equatorial air is warm and moist. That's why you find rain forests in the equatorial regions.

So how can there be snow at the equator? Simply because the weather stew is also affected by altitude. Although it is hot at sea level at the equator, it can get very cold as you move up into mountains. Some equatorial mountains, such as Mount Kilimanjaro in East Africa and the Andean peaks in Ecuador, are snow-covered year-round.

If the Equator Is So Hot, Why Are There No Deserts on the Equator?

So the equator gets the most sunshine. That means it should have the most deserts, right?

Many people hear the word *desert* and picture an old movie about the French Foreign Legion. A soldier in tattered khakis, lips cracked and parched, drags himself slowly through the dunes, croaking "Water." Or we conjure up cartoons of people stuck in the desert, envisioning unlikely mirages of soda fountains and waterfalls.

While most of us might equate the word "desert" with "hot," that isn't an accurate association. Technically speaking, a desert is an almost barren tract of land in which rain or other forms of precipitation are so scanty or irregular that most vegetation won't grow, except for extremely poor grass and low brush. Specifically, a desert is an area that receives less than ten inches of precipitation a year. In other words, deserts may not be hot, but they are dry.

Although there are low-lying "desert" islands along the equator where rain-bearing winds simply pass over, the major desert zones lie in two belts on either side of the equator. The warm, moist equatorial air that keeps the rain forests wet, flows toward the poles. As the air rises, it cools and dries. By the time it reaches the latitude of 30°—both north or south of the equator—this cool, dry air begins to sink, warm up, and flow back toward the equator. These dry cells of air are responsible for the belts of desert and arid land found around the Tropics of Cancer and Capricorn.

The world's deserts can be divided into subtropical deserts—such as the Sahara and Arabian deserts—and mid-latitude deserts such as the Gobi in Mongolia. Less familiar is the notion of a *cold desert* where water is unavailable because it is trapped in the form of ice. This means that the snowy vistas of Antarctica and Greenland actually qualify as deserts.

Whether too cold or too hot, high in the mountains or below sea level, sandy or rocky, all deserts are linked by conditions highly unwelcome to life. Aridity and extreme climates make farming and raising animals all but impossible except with sophisticated irrigation systems. Sparsely populated and seemingly remote, the world's deserts actually take up one third of the earth's land surface. What's more, they are growing, with fearsome consequences for humanity.

The World's Great Deserts

The Sahara (Arabic for "wilderness") is by far the world's largest, most desolate, and hottest desert. The world's highest recorded temperature was 136.4° F (58°C), recorded in Al'Aziziya, Libya. In the eastern Sahara, near Cairo, the sun shines an average of eleven hours and forty-seven minutes per day. The Sahara stretches across North Africa from the Atlantic Ocean to the Red Sea, covering an area of more than 3.5 million square miles, dwarfing the other great deserts of the world. Strange as it may seem, the Sahara was once an expanse of grassland that supported the kind of animal life associated with the African plains. Cave paintings found in the Sahara show elephants and giraffes and depict cattle herders roaming a greener Sahara.

Long thought of as a wasteland, the Sahara changed in significance with the discovery of oil, gas, and iron-ore deposits after World War II. The Sahara is a vast system that includes the **Libyan Desert**, which stretches over Libya, Egypt, and Sudan; the **Nubian Desert** in northeastern Sudan, stretching from the Nile to the Red Sea and once home of the ancient empire of Cush, which overran Egypt in 750 BC; and the **Arabian Desert,** which runs from the Nile to the Red Sea in Egypt.

The **Kalahari** is Africa's other major desert area, but it is significantly different from the Sahara. With an area of about one hundred twenty thousand square miles, the Kalahari spreads over the countries of Botswana, Namibia, and South Africa. Unlike the Sahara, the Kalahari is almost entirely covered with grass and woodland, with bare sandy stretches appearing only in the driest section in the southwest, where rainfall is very low. Although the Kalahari is home to some game, it is only sparsely populated by the nomadic Bushmen.

A smaller African desert is the **Namib**, located on the Atlantic coast of Africa in Namibia. It is the home of the world's tallest sand dune, a crescent-shaped "barchan" dune.

The Red Sea separates the North African desert system from the other great desert belts of the Middle East. Across the Suez from Egypt is the **Sinai Peninsula**, a patch of desert historically important as both an ancient battlefield and a modern one. It was the scene of the forty years spent in the wilderness by Moses and the tribes of Israel, and the location of Mount Sinai, the holy mountain on which Moses received the Ten Commandments, which is believed to be either of two existing Sinai peaks, Jebel Serbal or Jebel Musa.

Further East, in Saudi Arabia, there are two major desert areas, the **Nafud** (Red Desert) and the **Rub al-Khali** (Arabic for "Empty Quarter"), which are some two hundred fifty thousand square miles of dry, hot, windswept dunes. But two overriding factors have made this sparsely settled and parched feudal monarchy one of the world's most significant places. Beneath the deserts lie the world's largest petroleum reserves. And the Saudi Arabian desert is the birthplace of Islam (Arabic for "submission to God"), making the desert kingdom the center of the Islamic world.

The Middle Eastern deserts continue to the north with the **Syrian** or **Al Hamad Desert**, which covers the eastern region of Syria and extends into Iraq. Two other large desert systems dominate Iran, Iraq's eastern neighbor. They are the **Dasht-e-Kavir** (Great Salt Desert) and the **Dasht-e-Lut** (Great Sand Desert).

To the north of Iran is the **Kara Kum**, a desert of more than one hundred thousand square miles covering nearly 80 percent of Turkmenistan, and the nearby **Kyzyl Kum**, another desert covering more than one hundred thousand square miles in the republics of Uzbekistan and Kazakhstan.

Moving further east, beyond the Tien Shan Mountains that create the natural border between Russia and China, is the **Takla Makan**, a one-hundred-thousand-square-mile desert in the great basin between the Tien Shan range and the Altun Shan and Kunlun Shan ranges that rise into Tibet. The other large desert area of East Asia is the **Gobi** in Mongolia and northern China, the world's second-largest desert, covering more than five hundred thousand square miles. Long home to the wandering Mongols who once created a great empire that spread out of China and stretched from Vietnam to Poland, the Gobi has proven a vast storehouse of prehistoric finds. It is the site of significant fossil dinosaur discoveries, as well as many early human implements that date as far back as a hundred thousand years.

Most people think of steamy rain forests or teeming cities like Calcutta when they think of India. But there is a sizable desert, called the **Great Indian** or **Thar**, that straddles northwestern India and eastern Pakistan.

There is less land in the southern hemisphere and consequently fewer deserts. But about one half of Australia is a desert plateau. Although it is called the **Great Australian Desert,** Australia's large, arid interior section is actually made up of four separate deserts. These Australian deserts are the **Simpson**, the **Great Victoria**, the **Great Sandy**, and the **Gibson.**

In South America, Chile's **Atacama Desert** is very small. It is hemmed in by the Andes, which keeps out winds and any rain. The Atacama has the distinction of being the driest place on Earth. In 1971, it received rain for the first time in four hundred years.

In contrast to all of the world's great deserts, the deserts of the United States are not very impressive. The deserts of the American Southwest are also rather small when taken individually. But they are more striking when seen as part of a region known as the **Great Basin**, a vast area stretching from lower Oregon and Idaho, covering most of Nevada, western Utah, and down to southeastern California. The area is kept dry by the presence of the Rocky Mountains, which prevent warm, wet air from the Pacific from reaching the central states that lie within the "rain shadow" of the Rockies. The **Mojave** is the largest of America's deserts, covering fifteen thousand square miles. It is an area of barren mountains and desert valleys in Southern California that receives only five inches of rain per year. A rich source of minerals, the Mojave is also the location of **Death Valley**, the lowest point in the Western Hemisphere (282 feet below sea level).

The **Colorado Desert** lies south of the Mojave in California and extends into Mexico's Baja California. It is bordered on the east by the Colorado River and features the Salton Sea, a shallow saline lake that was created when the Colorado flooded in 1905.

The **Great Salt Lake Desert**, in Utah, extends west from Great Salt Lake to the Nevada border. The level part near the Nevada border forms the famed Bonneville Salt Flats, where land speed records have frequently been set.

The **Black Rock** and **Smoke Creek** deserts are in northwestern Nevada near the border with California.

The other famous American desert that is not part of the Great Basin is Arizona's **Painted Desert**, a brightly colored region of mesas and plateaus. Here, centuries of erosion have exposed red, brown, and purple rock surfaces that give the area its name.

The world's deserts are spreading, extending over more and more land. When the desert encroaches because of human activity, the process is called *desertification*. People have been creating deserts since the beginning of settled agriculture, ten thousand years ago. The once-fertile crescent lying between the Tigris and the Euphrates Rivers is now mostly desert. Yet it once supported tens of thousands of people and nurtured some of the world's first true cities. Centuries of overuse, combined with poor irrigation techniques, sterilized the land and was one of the main causes of the collapse of some of these early civilizations.

The human factors in desertification are simply human attempts at survival. Marginal lands are cleared and plowed in often futile attempts to plant crops, and trees are removed. Trees and woody plants are also slashed for fuel, adding to the problems of deforestation. The trees might block the wind that carries away topsoil. Faulty irrigation sterilizes the earth by introducing salts and alkalis into already depleted soils. Goats, sheep, cattle, and camels are allowed to overgraze, removing (and flattening) vegetation that would help keep soil in place.

By some estimates, the expense of rehabilitating the degraded lands, and of halting the spread of deserts, need be no more than $2.5

billion a year. That is a fraction of the cost of lost agricultural lands and their yields. But funds provided for these areas are increasingly used for emergency measures. Ironically, the remedies are not overly exotic, rather as simple as planting trees or preventing overcutting. It has already worked. Deserts are being turned back.

In northern China, a green wall of trees has been planted to hold off the advancing desert and to stabilize badly eroded uplands. In Rajasthan, India, imported acacia trees from the Middle East were used to stabilize sixty thousand acres of sand dunes. In Haiti, 35 million trees have been planted. And in parts of West Africa, selected trees have been used to revitalize exhausted cropland and pasture. The trees have leaves during the dry season, giving shade when it is most needed. They also act as a windscreen and transfer nitrogen from the air into the soil, increasing crop yields. The trees' pods and seeds provide protein-rich fodder for cattle and goats. In Kenya, the Green Belt Movement, started in the 1970s, has created hundreds of tree nurseries. Farmers—mostly women—have planted more than seven million trees around croplands to reduce wind erosion. In Niger, where wind is the chief cause of soil erosion, another project has planted rows of trees, creating a windbreak now over two hundred miles long. There are even simpler methods. In poor Burkina Faso, in northwestern Africa, farmers use an ancient technique. Lines of stones are placed in the fields, trapping soil that would otherwise have been washed away, and acting as dams to hold rainwater. Crop yields there have been raised by 50 percent.

Where Was the Dust Bowl?

Americans savor their football with the same passion that the people of most other nations reserve for their football, or soccer, as it's called in the United States. They watch the Super Bowl, the Rose Bowl, the Orange Bowl, the Cotton Bowl, and the Peach, Liberty, Hula, and Blockbuster bowls. They remember the winners and losers, the tackles and fumbles, and who scored the winning touchdowns. But far fewer Americans recall the Dust Bowl. It wasn't a game and nobody won.

For those who think desertification is an exotic thing that only happens someplace else, there is a stark lesson in the experience of the American Great Plains region in the 1930s, the period of the Great Depression. During the prosperous years in the 1920s, farmers expanded their holdings, plowing up large areas of grassland, digging more wells, and allowing their cattle to range over wider areas. Drawing on so much underground water dried the land and accelerated the erosion of some of the world's richest topsoil. When a major drought struck, the remaining soil was blown away by high winds in massive dust storms.

In a few years' time, by in the mid-1930s, what had been a thriving agricultural region was a dry, windblown desert. Hundreds of thousands of farmers were uprooted in the largest forced migration in American history. Families in Kansas, Oklahoma, Texas, Colorado, Nebraska, and other Plains states left behind worthless farmland in scenes captured by John Steinbeck in his classic novel, *The Grapes of Wrath* (1939).

Which Is Colder, Antarctica or the Arctic Circle?

Too frigid for humans without the greatest of precautions, hard to reach, impassably ice-bound for most of the year, the separate polar environments used to be among the least known places on earth. We knew more about the moon than we did about the two poles. But things are changing. With energy and mineral companies licking their chops as they eye the natural resources that might be trapped beneath the icy wastelands, the scene is set. The familiar story of conflict between exploitation of potentially valuable assets and preservation of these areas may dominate their future. For now, science has the upper hand as most exploration taking place at the two poles is largely of a research nature rather than a treasure hunt for new mineral wealth. But that could always change.

People tend to think of the two polar zones as being similar. But they are quite different. At the North Pole, the Arctic is the smallest of the oceans. There is no evidence yet of any native population living at the North Pole. But that isn't to say there isn't any Santa Claus. Just because you can't see him doesn't mean he isn't there.

Antarctica is an ice-covered land mass twice the size of Western Europe and surrounded by a vast open ocean. (See in Chapter 2, page 90, "Who Owns Antarctica?") The world's record cold temperature was the reading of –126.9°F (88.3°C) recorded at the Soviet Union's Vostok station in Antarctica. An average warm summer day in Antarctica is about 1°F. The average daily temperature of the six-month Antarctic winter is –70°F.

Land tends to be colder than water because of its density, so Antarctica is generally colder than the Arctic Ocean because it is a land mass, while the Arctic is water. But seasonal differences between the regions are typical. On average, the polar region of Antarctica during July is the coldest place on earth. But in January, the coldest temperatures occur in Siberia, Canada, and Greenland, hundreds of miles to the south of the North Pole.

Who Is El Niño?

Short for "El Niño de Navidad," or the Christ Child, El Niño is actually a "what" rather than a "who." A huge pool of extra-warm seawater that flows in a tonguelike current out of the tropical eastern Pacific Ocean, El Niño seems to occur unusually strongly about once a decade, usually in December, before Christmas, which is how he got his name. But El Niño does not come bearing gifts. Instead, this regular visitor brings climate havoc and is responsible for a big falloff in the anchovy haul, pushing up the price of pizzas. Its effects are not funny, however, and can be extremely destructive.

In 1982–83, El Niño was unusually dramatic, a whole 7°C warmer than normal water temperature ranges, pumping more heat energy into the atmosphere. A period of unusual weather at the time has been called the most disastrous in recorded history. Weather patterns were altered across three quarters of the globe, causing floods along the western coasts of both South and North America, and droughts in southern Africa, South Asia, and Australia. There were massive die-offs of fish, sea birds, and corals, and the human toll of El Niño was put at more than a thousand dead.

The 1990s edition seemed kinder and gentler. But it is still exerting strange effects on the world's weather. Again, some places were more rainy than usual. In others, drought conditions worsened. Although it was first recognized in 1726, El Niño is still little understood and no single theory for the once-in-a-decade warming is widely accepted.

NAMES: *Cyclone, Hurricane, Typhoon, Tornado*

"It's a twister!" At least it was in *The Wizard of Oz*. But what was Dorothy Gale's Kansas "twister"—a tornado or a cyclone? And what are the differences anyway?

By any name, these storms can be deadly. In April 1991, a cyclone in Bangladesh left 138,868 dead and millions homeless. Hurricane Gilbert, which hit the Caribbean and the Gulf coast of the United States in 1988, killed at least 260 people. In 1984, Typhoon Ike hit the Philippines, and more than 1,300 people died. A 1974 tornado killed more than 300 people in the American Midwest.

All four types of storms are variations on a true cyclone, a word that is used in both a specific and a general sense. A true cyclone is technically a windstorm with violent, whirling winds rapidly circulating around a low-pressure center—the "eye" of the storm. Cyclones are accompanied by stormy, often destructive weather. (In the Northern Hemisphere, the cyclone winds circulate counterclockwise; in the Southern Hemisphere they move in a clockwise direction.)

Hurricanes, typhoons, and cyclones are all the same kind of violent storm originating over warm ocean waters and called by different names all over the world. *Hurricane* (derived from a Carib Indian word) is typically used to describe these storms when they originate in the tropical Atlantic Ocean or Caribbean Sea. The huge whirling mass of cloud which rotates around the calm "eye" may be over 248 miles (400 km) across. The spiraling winds may reach from 9 to 24 miles (15 to 20 km) up into the atmosphere. A *typhoon* (from the Cantonese word *tai fung*) is a violent tropical cyclone originating in the western Pacific, and especially the South China Sea. When these storms originate in the area of the Indian Ocean, they are called *cyclones*.

A *tornado* (an alteration of the Spanish word *tronada*, for "thunderstorm") is also a form of cyclone. A violent twisting funnel of cloud that extends down to land from a large storm cloud, a tornado covers a much smaller area than a hurricane. But it is often more violent as it rushes across the land at speeds of 18 to 40 miles (30 to 65 kilometers) per hour.

What Is a Monsoon?

Monsoon comes from the Arab word *mausim*, which means "season." The word is used for the heavy rain that occurs in some parts of the world, especially in South and East Asia. This rainy season begins suddenly when winds from the sea sweep across the land. More than half the world's population lives in areas with monsoon climates. And most of these are the poor regions in which most of the people depend on agriculture for survival. A late or failed monsoon season can mean disaster and starvation.

A typical monsoon cycle occurs in India. There, from April to June, the weather gets hotter and hotter, and the air is very dry. Winds blow outward from the land. Then in June or July, the wind reverses direction and the monsoon bursts. The day when the rain will begin can often be forecast. Then winds begin to blow from the sea, bringing with them rain.

What's More Likely, Another Ice Age or a Glacial Meltdown?

Geologists are just beginning to scratch the surface of the ice when it comes to understanding long-range historical climate changes. Deep-drilling research currently going on in Greenland, for instance, will provide more evidence about the earth's historical temperature changes than any past research has offered to date. We do know that there have been regular warm periods between the extensive glacial eras or ice ages. During the colder ice-age periods, glaciers and vast snow sheets extended over large areas of the earth, changing sea levels and affecting the patterns of ocean and air currents.

Currently, the earth is enjoying a sort of "tropical vacation," a warm interglacial period. Historically, these warm periods have lasted about ten thousand years. But our current warm phase is about that old, and there are researchers who believe conditions are now similar to those of about ninety thousand years ago when a sudden period of cooling took place. Some climatologists even suggest that there is evidence of glacial advance during the past thirty years. Of course, the difference between past and present—and it is a big one—is the human factor. The human impact on the earth's long-term climate is a bit of a coin toss. Some scientists suggest that the dust and fumes we have been adding to the air since the beginning of the Industrial Revolution may create a "sun screen" or umbrella, that will reduce the sun's heat, lowering the earth's temperature and hastening the next ice age.

On the other side—which is probably more crowded—are the scientists who argue that just the opposite is happening. As human activities throw increasing amounts of certain gases into the atmosphere, a "greenhouse" effect is created that will effectively *raise* world temperatures, referred to as *global warming*. A lot of people don't take the greenhouse effect very seriously. Other skeptics are people who have heard scientists cry "The sky is falling!" one too many times. Maybe the problem is with the name "greenhouse." It sounds a little too attractive, too positive. After all, greenhouses are places where flowers and vegetables grow all the time. That doesn't seem so terrible.

Maybe it should be called the car-with-the-windows-closed-on-a-hot-day-at-the-beach effect. Maybe then people would take the threat of global warming seriously. After all, not many people have been inside greenhouses. But a great many more understand what it means to leave the car locked up tight in the sunshine on a hot day. Open the car doors and you get a furnacelike blast of superheated air in the face, the steering wheel is too hot to touch, and the upholstery practically sears your flesh.

Carbon dioxide is what makes seltzer and soda fizzy. Sounds pretty benign. It is also one of the most important gases in the atmosphere, even though it amounts to a mere .03 percent of it. When the sun's energy strikes the earth's atmosphere, much of it bounces back into space. But some is absorbed by carbon dioxide, warming the surface of our globe through what is commonly referred to as a greenhouse effect. Carbon dioxide and other gases in the atmosphere act like the glass in a greenhouse—or your car windows—letting the sun's rays through but trapping some of the heat that would otherwise be radiated back into space. Humanity needs a certain level of carbon dioxide to make life possible. The average temperature of the planet might be 30°C colder if not for the carbon dioxide factor.

This is not a new idea. In 1896, a Swedish chemist named Svante Arrhenius coined the term "greenhouse effect" and predicted that the burning of fossil fuels would increase the amount of carbon dioxide in the atmosphere and lead to a warming of the world's climate. Arrhenius was pretty smart—and about a hundred years ahead of his time.

Scientists are now broadly agreed that the greenhouse effect is bringing about the greatest and most rapid climate change in world history. It will have enormous consequences for life on earth. Since fossil fuels began to be used on a massive scale with the onset of the Industrial Revolution, huge amounts of carbon dioxide have been released, and this trend is now being aggravated by the burning of our tropical forests. About four fifths of the carbon dioxide now comes from the burning of fossil fuels. The rest is from destroying vegetation, mainly the felling of forests.

Why save the rain forests? Trees soak up carbon dioxide when alive, but release it when they are cut down and burned.

The result is a steady warming of our planet. While there may be little change at the equator, the poles may well become 7°C warmer, with all the ultimate implications for the ice caps. As the polar ice caps could eventually melt, sea levels might rise by five to seven meters (sixteen to twenty-two feet). Much of the Netherlands would be flooded, together with half of Florida and huge sectors of other low-lying areas such as Bengal, and many low-lying Pacific islands where the rise in sea levels will diminish freshwater supplies even before flooding becomes a problem. Even a one-meter rise in the sea level could make 200 million people homeless.

But even this is likely to be overshadowed by the impact of global warming on harvests. While a warmer world might be able to grow more food overall, some nations will be winners and others losers. The American Midwest, which helps to feed a hundred nations, may see its harvests cut by about a third. The United States will still be able to feed itself, but exports to the rest of the world could fall by up to 70 percent. New land will open up in Canada as the weather warms, but the soils are too poor to make up the loss. The optimistic view is that Ukraine, once the bread-basket of the Soviet Union, may not be hit so badly, and the land that opens up in Siberia is better than that in Canada. But there is a darker forecast that holds that the ice concentrations under the frozen Siberian tundra would make the land useless for farming if it defrosted. There is also some concern that methane, another dangerous greenhouse gas, is trapped in the frozen tundra and might be released, with even more harmful results. A 2009 study, reported in the English *Daily Telegraph*, said the melting permafrost could trigger "unstoppable climate change" as the gases released would contribute further to the warming.* Of course, poor countries will be hardest hit. Areas that are already arid— like Tunisia, Algeria, Morocco, Ethiopia, Somalia, Botswana, eastern Brazil, and parts of Asia—will probably dry out even further.

The 1980s were by far the hottest decade ever recorded. Seven out of the ten warmest years recorded since 1880 occurred during this decade. And 1990 was ranked in several studies as the warmest year yet. No one can be sure whether this is due to the greenhouse effect or whether it was simply the result of natural variability in the climate. A slight cooling took place in 1991, but that was attributed to the cloud of dust thrown up by the eruption of Mount Pinatubo in the Philippines and would most likely be a temporary situation slightly delaying an overall warming trend.

Geographic Voices   Elisabeth Rosenthal and Andrew Revkin, *New York Times*, February 3, 2007

In a grim and powerful assessment of the future of the planet, the leading international network of climate scientists

---

* Louise Gray, "Melting Permafrost Could Trigger 'Unstoppable' Climate Change," *Daily Telegraph*, March 26, 2009.

has concluded for the first time that global warming is "unequiv-ocal" and that human activity is the main driver, "very likely" causing most of the rise in temperatures since 1950.

Is All the Talk of Global Warming Just Hot Air?

So it comes down to this. Is it happening? If you read the scientists on this question, the very large consensus is, Yes, the earth's climate is warming and the warming has accelerated.

Most Americans don't bother to actually read the science, though, and rely on the mass media, which always has to make a claim of being balanced by presenting two sides of the story, even when there isn't re-ally a credible "other" side.

The other problem is that people listen to politicians or others with agendas not driven by science. As Naomi Oreskes, author of *Merchants of Doubt*, wrote:

> Politicians, economists, journalists and others may have the impres-sion of confusion, disagreement, or discord among climate scientists, but that impression is incorrect. The scientific consensus can be wrong. If the history of science teaches anything, it teaches humility, and no one can be faulted for failing to act on what is not known. But our grandchildren will surely blame us if they find that we understood the reality of anthropogenic climate change and failed to do anything about it.

Is it our fault? That's where it gets trickier. Again the very large consensus—90 percent by some estimates—says Yes, humans are re-sponsible for the activities that are causing the warming—primarily through the burning of fossil fuels and deforestation.

Achim Steiner, executive director of the United Nations Environ-ment Program, which administers the Intergovernmental Panel on Climate Change (IPCC) along with the World Meteorological Orga-nization, told the *New York Times*, "In our daily lives we all respond urgently to dangers that are much less likely than climate change to affect the future of our children. Feb. 2 [2007] will be remembered as the date when uncertainty was removed as to whether humans had anything to do with climate change on this planet. The evidence is on the table."

Veteran science writer William K. Stevens agreed, in a column entitled, "On the Climate Change Beat, Doubt Gives Way to Certainty." Stevens wrote in 2007:

> The Intergovernmental Panel on Climate Change said the likelihood was 90 percent to 99 percent that emissions of heat-trapping greenhouse gases like carbon dioxide, spewed from tailpipes and smokestacks, were the dominant cause of the observed warming of the last 50 years. In the panel's parlance, this level of certainty is labeled 'very likely.' Only rarely does scientific odds-making provide a more definite answer than that, at least in this branch of science.

So given the current state of professional knowledge and a large and growing body of research and data, there is little scientific debate over the reality of global warming, and only slightly less disagreement in the scientific world over what role humans have played in creating the current crisis by burning fossil fuels and deforestation. But there are scientists who say the greenhouse effect has been greatly exaggerated.

Yet most scientific groups are now quite sure that the greenhouse effect is happening. Among them is the Geological Society of America, which issued an amended statement in 2010 reading:

> Decades of scientific research have shown that climate can change from both natural and anthropogenic causes. The Geological Society of America (GSA) concurs with assessments by the National Academies of Science (2005), the National Research Council (2006), and the Intergovernmental Panel on Climate Change (IPCC, 2007) that global climate has warmed and that human activities (mainly greenhouse gas emissions) account for most of the warming since the middle 1900s. If current trends continue, the projected increase in global temperature by the end of the twenty-first century will result in large impacts on humans and other species. Addressing the challenges posed by climate change will require a combination of adaptation to the

changes that are likely to occur and global reductions of CO_2 emissions from anthropogenic sources.

Among the last serious doubters on the question of the human role—"anthropogenic" means caused by humans—was the American Association of Petroleum Geologists (AAPG). Apparently facing widespread dissent from many of its members, the AAPG also issued a revised 2010 statement that was more accepting of the human factor:

> The AAPG membership is divided on the degree of influence that anthropogenic CO_2 has on recent and potential global temperature increases . . . Certain climate simulation models predict that the warming trend will continue, as reported through NAS, AGU, AAAS and AMS. AAPG respects these scientific opinions but wants to add that the current climate warming projections could fall within well-documented natural variations in past climate and observed temperature data. These data do not necessarily support the maximum case scenarios forecast in some models.

Given the most optimistic estimates, global warming from the car-with-the-windows-closed effect is happening around us with potentially disastrous effects. The temperatures will not only continue to increase, with dangerous pollutants as well, but sea levels will continue to rise, further threatening low-lying coastal areas, especially in catastrophic weather conditions. That is significant because with rising temperatures, there is a growing belief that tropical storms are becoming more dangerous. The essential idea is that warmer ocean waters produce more dangerous tropical storms and hurricanes, which derive their energy from ocean waters.

The best the world can do now is park the car in a shady spot and roll down the windows. In all likelihood, with even radical changes in fuel usage and protection of forests, the best we can do is slow the process down rather than eliminate the threat of global warming.

It is a debate, however, that many governments have moved past. Many countries are preparing for a new era in which the iced-over Arctic will eventually melt, opening up a treasure trove of resources. In a report published by the Associated Press in 2012, Eric Talmadge reported that several countries, including the United States and Russia, that are near the Arctic are making serious military preparations to defend—or potentially fight over—the Arctic. The U.S. Geologic Survey estimate that 13 percent of the world's undiscovered oil and 30 percent of its untapped natural gas is in that region.

These are some of the expected consequences of a warmer world. And the fight for the world's resources on a much greater global scale may already have begun.

Geographic Voices   Jim Robbins, *"Why Trees Matter"*

We have underestimated the importance of trees. They are not merely pleasant sources of shade but a potentially major answer to some of our most pressing environmental problems. We take them for granted, but they are a near miracle. . . . What trees do is essential though often not obvious. Decades go, Katsuhiko Matsunaga, a marine chemist in Japan, discovered that when tree leaves decompose, they leach acids into the ocean that help fertilize plankton. When plankton thrive, so does the rest of the food chain. In a campaign called Forests Are Lovers of the Sea, fishermen have replanted forests along coasts and rivers to bring back fish and oyster stocks and they have returned.

The "Antarctic Donut": Powdered or Jelly?

One of the other nasty little chemicals we have been spewing out into the atmosphere is a compund called a *chlorofluorocarbon*, or CFC, and it is also considered a greenhouse gas along with carbon dioxide and methane. But another proven effect of CFCs is the "Antarctic donut," a huge hole in the atmosphere's ozone layer over the South Pole. This ozone hole opens over Antarctica every southern spring. It is as wide as the United States and as deep as Mount Everest is tall. The discovery of the Antarctic donut in October 1982 caught many scientists by surprise. Computers on a satellite measured the hole, but scientists rejected the data as too unlikely. Recent data also suggest that another tear in the ozone fabric may be taking place in the Northern Hemisphere.

At first glance, the ozone layer doesn't sound very appealing. In fact, it is a shroud of poison enveloping the earth. A form of oxygen, ozone is highly toxic; less than one part per million of the blue-tinged gas in air is poisonous to humans. Near ground level, it is a pollutant that helps form smog. But far above, in the stratosphere, it forms a lifesaving screen on which all life depends. Ozone is the only gas in the atmosphere that can screen out the lethal ultraviolet rays of the sun. If it were not for this thin screen of ozone, ultraviolet radiation would kill all terrestrial life.

Small amounts of ultraviolet radiation do get through this fragile filter. It is the main cause of skin cancer, a rapidly increasing disease that already kills some twelve thousand people a year in the United States alone. Dermatologists in the United States are already seeing many more cases of skin cancers among teenaged and adolescent patients. Ultraviolet radiation suppresses the immune system, helping cancers to become established and grow, and increases susceptibility to other diseases. It is a major cause of cataracts, which blind at least 12 million people worldwide and damage the sight of at least another 18 million. This radiation also diminishes crop yields and kills ocean microorganisms that serve as food for other species. There is no question that damaging the ozone layer even slightly will increase the toll on human health.

The greatest danger to the ozone layer comes from CFCs—outstandingly useful and versatile chemicals. They were first developed as coolants and played an essential role in the implementation of air conditioning and refrigeration. They were then introduced as aerosol propellants. But every CFC molecule lives on to destroy thousands of molecules of ozone.

It will be a long time before the ozone returns. U.S. Environmental Protection Agency analyses suggest that, even if all ozone-depleting chemicals are phased out, it will take a century for conditions in the atmosphere to return to what they were in 1986. But the world did move quickly to agree to phase out the chemicals once their dangers were demonstrated. The DuPont Company, the largest producer of

CFCs, voluntarily pledged to phase out CFC production, probably an unprecedented example of corporate responsibility. Action on CFCs offers one of the first hopeful precedents for international action on other environmental threats.

We are just left to wonder if it is too little, too late.

What Is Acid Rain?

First described in 1872 by an English chemist, acid rain stands as one of the industrialized world's nastiest problems, the most controversial form of air pollution in the developed world. It is bitterly ironic that something we think of as so generally positive—gently falling rain—is like taking a shower in battery acid. Seemingly simple "rainfall" is degrading and destroying both nature and the achievements of humanity.

Acid rain, including acid sleet and snow, is produced primarily by the release of sulfur oxides into the atmosphere. The chief sources of such emissions are electrical generating plants, industrial boilers, and large smelters. Gases that are vented into the air by tall smokestacks get caught up in prevailing winds where, in the course of transport over land, they are transformed into dilute solutions of sulfuric acid and nitric acids. Ironically, part of the acid-rain problem comes from earlier attempts to clear local polluted air by raising smokestack heights. All that managed to do was dissipate the pollutants higher into the atmosphere, where they could blow into the neighbor's backyard; Canada's problems come from the United States and London's dirty air blows into Norway.

When these industrial pollutants combine with water vapor, sunlight, and oxygen in the atmosphere, they create a diluted soup of sulfuric and nitric acids. This appetizing mixture is then washed out of the atmosphere by rain, snow crystals, or in the form of dry particles. Eventually, it reaches the water cycle, increasing the acidity of freshwater lakes, streams, and soils. Nearly a quarter of Sweden's ninety thousand lakes are acidified to some extent. Fish can no longer survive in four thousand of them. In Norway, many lakes and streams are technically dead. And in the eastern United States, thousands of lakes are now too acidic to support fish. Trout and salmon no longer reproduce in nine acidic rivers in Nova Scotia.

In parts of Pennsylvania, the result is a corrosive solvent a thousand times as acidic as natural rain. So far, the worst hit by the acid rain nightmare are northeastern regions of Canada and the United States, Central Europe, and Scandinavia. But Australia and Brazil are also noticing early signs of the deadly rain.

Objects ranging from the Great Pyramids and the Sphinx to the great architectural landmarks of Europe are being eaten by acidic smogs. The masonry of Cologne (Köln) cathedral is being eaten away. Many of Europe's stained-glass windows are fading. In West Germany, the Black Forest has lost one third of its trees, and many scientists attribute this trend to a combination of acid rain and other forms of air pollution. It is difficult to assess the costs of acid-rain damage. But it has been estimated that damage to metals, buildings, and paint in European countries costs around $20 billion a year, and that does not include costs of dead forests, acidified lakes, and damaged crops. Damage to the West German timber industry alone is estimated at $800 million, plus an additional $600 million in agricultural losses due to reduced productivity.

Acid rain—with other human assistance in the form of logging, clearing of forests for cattle pasture, wood cutting for fuel by half the world's population—is killing off one of nature's best defenses of the atmosphere, the world's forests.

As they used to teach us in grade school, "forests are our friends." They play major roles in the planetary recycling of carbon, nitrogen, and oxygen—in other words, "in with the bad air, out with the good air." Forests help to determine temperature and rainfall. They are often at the source of great river systems and protect soil from eroding. They constitute the major gene reservoirs of our planet, and they are the main sites of the emergence of new species.

Wood was once one of the earth's most plentiful resources. But it has been treated badly. In 1950, 30 percent of the land was covered by forest, half of which was tropical forest. By 1975, the area covered by tropical forest had declined to 12 percent.

After a bitter fight with the Canadian government, the United States reluctantly acknowledged the problem and began taking action to reduce emissions. Large fossil-fuel power plants were supposed to cut emissions of sulfur dioxide by about 40 percent by 1988. Under the Clean Air Act of 1990, the United States created a new program called "cap and trade." It allowed power-plant operators to buy, sell, and trade credits to continue to pollute as long as they reduced overall emissions by half.

The program proved largely successful, as the *San Francisco Chronicle* reported in 2007. In 2000, the Environmental Protection Agency began the second phase, where nearly all power-generating units were forced to reduce their sulfur dioxide output or buy credits from plants that cut their emissions. Plants that miss the targets face stiff fines—two thousand dollars per excess ton. At the time the legislation was being debated, many environmental groups were suspicious of the idea of a pollution credits trading program. Some dubbed it a "license to pollute."*

While skeptics of the approach remain, the "cap and trade" program was deemed a success, one that could be applied to greenhouse gases as well—a larger and far more difficult problem than pollution.

What Are Wetlands?

This question is not as simple as it sounds. For instance, did you ever hear of the National Wetlands Coalition? Sounds like a nice ecologically correct organization, a gentle "save the earth" group devoted to protecting natural resources. In fact, this coalition includes representatives of five oil companies, and one of their missions is to get the United States government to define "wetlands" in a manner more to their liking.

During his presidential campaign of 1988, when George H. W. Bush pledged he would be the "environmental president" as well as the "education president" and the "war against drugs president," the

----

* www.sfgate.com/green/article/Cap-and-trade-model-eyed-for-cutting-greenhouse-3300270.php#ixzz2Fcd5hDNS

candidate also said that "all existing wetlands, no matter how small, should be preserved." This was no small promise. The United States currently has about 100 million acres of wetlands. But America has lost half its wetlands, most of them to agriculture, and largely during the past twenty years. Florida's Everglades, for instance, were progressively drained for agriculture for a century. But plenty of acres of wetlands have also been paved over to make way for condos and shopping malls.

And here's where things get tricky. How do you define "wetlands"? One person's wetland is another person's swamp, marsh, or bog. The world's wetland areas—which also include fens, estuaries, and tidal flats—make up 6 percent of the earth's land surface and are among the most fecund and productive ecosystems in the world. According to the World Wildlife Fund's *Atlas of the Environment*, "they provide critical habitats for thousands of species of plants and animals, yield up food, fiber and building materials, play important roles in regulating water cycles, filter pollution and guard shorelines from the depredations of the sea."

Coastal wetlands—estuaries, salt marshes, and tidal flats—are vital spawning and nursery areas for fish and shellfish. Two thirds of the fish caught worldwide are hatched in tidal zones. Wetlands also act as nature's pollution-control and sewage-treatment plants. Viruses, coliform bacteria (from fecal matter), and suspended solids normally left after waste is processed in secondary sewage treatment plants can be transformed and made harmless through wetlands. In Hungary, for instance, peat bogs have long been used as natural filters for waste water from sewage plants.

The threat to these valuable systems isn't limited to the United States. Wetlands along the Nile delta, the once incredibly fertile area where Western civilization was practically born, are shrinking at a fast pace. Irrigation projects and drainage of coastal wetlands are destroying an area that is home to 50 million Egyptians. Near southern Spain's Doñana National Park, a sort of European Everglades, conservationists are fighting developers who want to replace the important breeding grounds with hotels, marinas, and golf courses. Farmers want the land to be drained for cultivating rice and strawberries. Conservationists argue that an environmentally correct resort that would attract nature-loving tourists is preferable to the planned development. To the farmers, they recommend turning to crops that demand less water. The situation in southern Spain is a typical case of the battle between conservation and outright development. Named to the UN's World Heritage Site list, the park's general state of preservation in 2012 was called "satisfactory." But it continues to face numerous threats, including agricultural development, tourism, poaching, overgrazing, and illegal exploitation of crayfish.

According to UNESCO, "Doñana National Park has been a testing ground for conservation in Spain and has become very well known throughout Europe due to the controversies faced there and the innovative management approaches that have been taken . . . Doñana National Park is a resilient system, and nature is still the dominant force. As the main threats have been averted and as restoration activities are under way, the future of the park seems assured.*

Where Is the World's Most Populous City?

Towered cities please us then,
And the busy hum of men.

The poet Milton liked cities. But two centuries later, his countryman, Lord Byron, disagreed:

High mountains are a feeling, but the hum of human cities torture.

Lord Byron may have been feeling claustrophobic back in 1812. How about you?

- Global population, which stood at 2.5 billion only forty years ago, hit 7 billion in March 2012, according to the U.S. Census Bureau.

* http://whc.unesco.org/en/list/685

- Current projections are for global population to reach 10.5 billion by 2050.

- If Africa maintains its present 3 percent growth rate until this time next century, its current half billion people will increase to 9.5 billion.

- Before 1850, only London and Paris had populations of more than 1 million. (Historical reconstructions have produced estimated populations of more than 1 million for several ancient cities, however, all of these gradually declined. Before the Industrial Revolution, Rome, Xian in China, Baghdad, Byzantium, and Edo—modern Tokyo—may have reached the 1 million mark for a short period before suffering major population losses.) In the second half of the twentieth century, more than two hundred forty cities reached the million mark, with most of them in developing countries.

The world's large cities, a relatively recent phenomenon in human history, are growing too fast for their own good. Ancient Rome may have been the first city of a million people in the first century AD. But after a few centuries of barbarians, wild and crazy emperors, and catastrophic epidemics, its population fell to around twenty thousand. Rome didn't reach the million plateau again until 1930. Most of the world's largest cities have grown up since the middle of the fifteenth century, the legacy of a phenomenal expansion in European trade. Thousands of towns—most of them ports located on coasts or major rivers—were developed as base cities for the growing network of global trade.

Following the great trading era, the next burst came during the Industrial Revolution, as cities in rapidly industrializing Europe and North America sprang up around manufacturing and shipping centers. The decline of many of these old "smokestack cities" during the past few decades has seen the rapid growth of cities shift to other parts of the world.

World's Largest Cities

By 2011, "urban agglomerations"—whole metropolitan areas comprising an urban center and surrounding areas—continued their explosive growth, especially in China, India, and other parts of Asia. According to the UN Population Division, the largest urban areas in the world were as follows:

Urban area	2011	2025 (UN projection)
Tokyo, Japan	37,217,000	38,661,000
Delhi, India	22,563,600	32,935,000
Mexico City	20,445,800	24,580,900
New York City metro area, U.S.	20,351,000	23,572,000
Shanghai, China	20,207,600	28,403,000
São Paulo, Brazil	19,924,500	23,174,700
Mumbai, India	19,743,600	26,556,900
Dhaka, Bangladesh	15,390,900	22,906,300
Kolkata, India	14,402,300	18,711,000

SOURCE: *The World Almanac and Book of Facts 2013.*

Founded by the Jesuits in 1554 São Paulo in Brazil passed the 20 million mark in 1988. The city, now the most important manufacturing base in Latin America, grew rich on the coffee industry, which provided the capital for the city's massive industrial growth. The promise of jobs in those factories has drawn millions of peasants from the surrounding countryside. An estimated 10 percent of the city's people now live in its slums and squatter settlements, with water pollution and a rapidly rising infant mortality rate among the major problems. The industrialization has also created an acid rain problem in the city, with acid levels a thousand times that of normal water.

Also growing fast is Lagos, Nigeria. The capital city grew rapidly after oil revenues began drawing hundreds of thousands of Nigerians to the city in hopes of finding a better life. They still come at a rate of 200,000 each year, a population influx that is only worsening the city's unofficial distinction as being possibly the world's worst city. Impossible overcrowding simply continues to worsen the problems of any typical large city: pollution, poor sewage facilities, chronic water shortages, traffic, high unemployment, poor housing, food shortages, and substandard medical treatment. During the 1980s, the worldwide drop in oil prices only served to exacerbate the problem. Like Mexico City, Lagos is built on swampy, infilled land. Drainage canals meant to take away some of the water that provides breeding grounds for mosquitoes and other disease-carrying insects have simply been filled in and built over. Since 1976, the Nigerian government has been working on a plan to relocate the capital to Abuja, an inland city at the geographical center of the country. Abuja became the capital in 1991 and Nigeria moved its official offices to the new capital, but the shift—like the attempted shift of Brazil's capital from Rio de Janeiro to the new inland city of Brasilia—has failed to weaken the pull of Lagos's magnetic attraction.

Can cities work in the future? That is a billion-dollar question. The world's richest cities, from New York to Tokyo, face intractable problems that come from too many people crowded into too little space, all creating garbage and sewage and requiring tremendous quantities of fresh food and water to be brought in. There are interesting experiments in creating workable cities in a variety of places around the world. One that is being viewed as a model for cities in developing nations is Curitiba, a commercial and industrial center of the Brazilian state of Paraná. Although the population there has grown elevenfold in the past half century to a little more than a million, it is leading the way in low-cost solutions to problems typical of many large cities in developing countries, especially in Latin America, where 75 percent of the population now lives in overcrowded cities, leaving the rest of the vast continent nearly empty.

In Curitiba, ambitious urban planning has been the key. A system of express buses provides inexpensive mass transportation at a fraction of the cost of subways, while reducing automobile traffic in the city by 25 percent. Large sections of pedestrian areas are free from cars and are crowded with shoppers. Exclusive bicycle roads have been constructed and factories help their workers finance the purchase of bicycles, an idea borrowed from the Chinese. Instead of massive urban-development projects, the city has attempted to recycle its older buildings. City officials also recognize that jobs must be found for the country people who continue to flood into the city, and old buses have been converted into classrooms where basic job skills are taught. For the young, there are apprenticeship programs, providing schooling, meals, and a small amount of pay along with real skills. To battle the garbage problem in areas where collection services cannot reach, the poor are given food in exchange for sacks of trash and refuse brought to collection centers.

Henry Thoreau, the man who said we should "simplify," would be proud.

Can the World Feed Itself?

Even more pressing than the problems of urban overcrowding presented by the world population boom is the more basic question of having enough food to go around. Ethiopia's predicament captures the essence of famine across the African continent, and Africa exemplifies a worldwide problem. During the headline-making famines of 1985, 35 million Africans suffered from acute hunger. According to some estimates, 150 million Africans suffer some degree of hunger and malnutrition even in non-drought years.

The simplistic answer to the question of world hunger is "Yes, there is enough to go around." But then it gets tricky.

William Clark, president of the International Institute for Environment and Development, recently wrote:

> Many detailed scientific studies have shown that the entire planet's population can be adequately housed and fed and provided with a livelihood which allows them to live beyond the fear of poverty. There are even sufficient resources for the six or more billion people who will be here by the end of the century. The issue is how existing resources are managed. The key to the future is the concept of sustainable development.
>
> By sustainable development I mean the rational use of resources to meet all basic human needs. To be sustainable, development cannot

ignore long-term costs for short-term gains. Concern for the environment is not a luxury that only richer nations can afford. If some development project is damaging forests or soil or water or clean air, then it is not true development.

Clark and other experts in the field of world agriculture and food development agree. The undernourishment of some 500 million people today does not stem from a global scarcity of resources. There is no doubt that the world could produce enough food. Yet tens of millions starve, and millions more are malnourished, because nature hasn't been "fair" in allocating resources. But this is where things get really tricky. National and global economic reforms are needed. But once you start to talk about the inequities of the distribution of the earth's resources, the haves get mighty nervous about the have-nots, and any talk of the "redistribution of wealth" sends shivers down the spines of the people whose wealth is most likely to be redistributed.

But there are some basic reasonable and sensible adjustments that seemingly must be made if the world wants to ensure that tens of thousands of babies won't die each day from diseases exacerbated by malnutrition. Improved agricultural techniques have already increased crop production radically in many areas of the world that once could not support themselves without importing basic foods. Simply upgrading tools from the Stone Age implements still used in some parts of Africa would help increase crop yields. Appropriate irrigation and use of fertilizers, the planting of the right crops, and reforesting to lessen the effects of drought and desertification are all simple and relatively inexpensive responses.

If peace holds in this ethnically torn nation and if the peasants are given proper encouragement, Ethiopia could feed itself one day.

One big problem the world over—watch out Ronald McDonald—is that about 40 percent of the world's grain goes to feed livestock. In the United States that figure is 90 percent! It takes at least ten calories of feed to produce one calorie of steak—an absurdly inefficient way to make food. And that is not to mention the large amounts of fresh water needed to raise cattle—very thirsty animals in a world running low at the water tank.

This is not a call for the elimination of our four-legged friends. Most domesticated animals around the world are quite efficient. They forage off plants that humans don't use—and they do it with no adverse consequences for natural environments, except when their numbers rise to unsustainable levels. But a simple moderation—not elimination—in the consumption of beef could have major consequences. Besides that, even the United States government has finally caught up with good medical sense and called for Americans to reduce their red-meat intake. So it's not only good for the rest of the world, it's good for you too!

Milestones in Geography VII
1992–Present

1992

Hurricane Andrew, a Category 5 hurricane, kills sixty-five people and causes $26 billion in damage to Florida and other U.S. Gulf Coast areas.

1993 A truck bomb planted by terrorists explodes under the World Trade Center in New York, killing six people.

The "Storm of the Century" strikes the U.S. East Coast, with massive flooding along the Mississippi and Missouri Rivers, causing $60 billion in damages and loss of power for 10 million.

1994

The North American Free Trade Agreement (NAFTA) goes into effect.

The Northridge earthquake kills seventy-two in the Los Angeles area.

1995

A heat wave kills seven hundred fifty in Chicago, bringing attention to the plight of urban poor and elderly in extreme weather.

The Khobar Towers bombing by terrorists in Saudi Arabia kills nineteen U.S. servicemen.

1997

Global economic crisis leads to a major stock market decline.

1998

U.S. embassies in Tanzania and Kenya are bombed by terrorists, killing 224 people.

1999

The world prepares for the possible effects of the "Y2K bug," a computer-related problem that was expected to cause computers to become inoperable.

2000

The USS *Cole* is bombed by terrorists in Yemeni waters, killing seventeen U.S. Navy sailors.

2001

September 11 terrorist attacks in New York City and Washington, DC.

United States launches invasion of Afghanistan.

2003

The space shuttle *Columbia* disintegrates upon re-entry to the earth's atmosphere.

The United States, Great Britain, and other allies invade Iraq.

2004

Four deadly and damaging hurricanes impact Florida, killing one hundred people.

2005

Hurricane Katrina devastates Louisiana, Mississippi, and Alabama coastlines. It is the most deadly and costly disaster in U.S. history, killing at least 1,836 people and causing $81 billion in damage.

2008

An outbreak of tornadoes kills more than sixty people in Arkansas, Kentucky, Tennessee, and Alabama.

Hurricane Ike kills one hundred people along the Texas coast.

The global financial crisis strikes.

2010

The Deepwater Horizon oil rig in the Gulf of Mexico explodes, resulting in the worst oil spill in U.S. history.

2011

The "Arab Spring," a wave of political upheaval, sweeps across North Africa and the Middle East, bringing new government to at least four countries, including Libya and Egypt.

The United States joins a UN military intervention in Libya's civil war.

2012

"Superstorm Sandy," a hurricane that formed in late October, devastated communities in the Caribbean and in several Mid-Atlantic and Northeastern states and crippled much of the New York City metropolitan area for weeks. At least 253 people were killed in seven countries along the path of the storm, which was estimated to be the second-most costly Atlantic hurricane after Katrina; damage to property in the United States was estimated at more than $63 billion.

Geographic Voices   Joel Kotkin, from *The Next Hundred Million: America in 2050*

The addition of a hundred million more residents also will place new stresses on the environment, challenging the country to build homes, communities, and businesses that can sustain an expanding and ever-more-diverse society. America will inevitably become a more complex, crowded, and competitive place, highly dependent, as it has been throughout its history, on its people's innovative and entrepreneurial spirit.

6

Lost in Space?

Astronomy compels the soul to look upwards and leads us from this world to another.

—Plato
The Republic

This is the excellent foppery of the world, that, when we are sick in fortune . . . we make guilty of our own disasters the sun, the moon, the stars; as if we were villains by necessity, fools by heavenly compulsion, knaves, thieves and treacherers by an enforced obedience of planetary influence.

—Shakespeare
King Lear

This is a new ocean, and I believe the United States must sail upon it.
—President John F. Kennedy

How Big Is the Universe?

How Far Is a Light Year?

Was There a Big Bang?

Did an Asteroid Kill the Dinosaurs?

Milestones in Space Exploration

Agentle warning to readers. Be careful here. If you were surprised to learn how big the oceans are, thinking about space may leave your head spinning. Once you start asking questions about outer space and our place in it, you realize how tremendously insignificant the earth is in the unimaginable vastness of the universe. Such questions have been lumped together and called *cosmology*, the curious intersection of science, faith, cold reason, and metaphysical musings.

But why discuss space at all in a book about the earth's geography? First, because most of humanity's conceptions of the world came about through observations of the heavens, which is why astronomy has been called the first science. Understanding the seasons, developing the calendar, and navigation all resulted from celestial observations made thousands of years ago.

In the second place, geography is about exploration and discovery. People have always asked, "What's over that mountain?" and "Who lives on the other side of that lake?" Even if the motives have often been greed or the desire to dominate the next-door neighbors, there is no denying that wanting to know about the places we can't see has been the prime mover throughout history. With fewer unexplored places left on earth, moving beyond the bounds of earth's gravity represents the last and greatest "frontier" and perhaps the most potentially rewarding realm of undiscovered riches.

And finally, if geography is simply about where we are in the world, it makes sense to understand where our world is in relation to all those heavenly bodies that we see at night, and the big one we see by day that keeps the earth alive—the sun. The secrets held by the universe about the creation of the earth, the beginnings of life, and the fate of the earth are being examined as we look to space to see where humanity will head next, and to seek an answer to one of the Really Big Questions: "Are we alone?"

How Big Is the Universe?

While astronomy has rightfully been called the first science, it might also be safe to call it the worst science. But it is not for lack of trying or general stupidity on the part of astronomers. On the contrary, it is astonishing to realize how *right* astronomers have been, given the tools they had to work with. For centuries, it was like using a kitchen knife to do brain surgery, and with the lights turned off.

Until fairly recently, astronomers were the scientific equivalent of the proverbial six blind men trying to describe an elephant by feeling different parts of its body. One felt its tail and thought it was a snake. Another grasped its leg and said it was a tree trunk. They just couldn't get the whole picture, so they were left to grope around the separate parts and imagine what the rest of the elephant looked like.

But a burst of extraordinary discoveries during the 1980s radically transformed our notions of the universe. It's not that today's astronomers have been given smart pills. They've just been given better tools— or more specifically, better eyes. It is a little daunting to imagine what Galileo, Kepler, Newton, or Einstein might have come up with if they had had the information and the technical wizardry astronomy has in its workshop today. And in the new century, a new generation of space-based and earth-based instruments, capable of seeing things science could never see before, will provide some of the missing pieces in the immense jigsaw puzzle called the universe.

One of the most basic notions the astronomically untutored must contend with is the idea that looking at space means looking at time. The light we see in the stars of deep space is light emitted long, long ago. In that ancient light are the clues that science hopes will answer many of the most basic questions people have asked since they started to crane their necks upward at the sun and the stars.

Ancient ideas about the size and shape of the universe coalesced in the writings of Ptolemy, whose notion of an earth-centered universe lasted for some fourteen hundred years. This notion was gradually challenged, first by Copernicus and then nudged along by Tycho Brahe and Johannes Kepler. Finally, Galileo raised papal eyebrows by stating that the earth was not the center of the universe at all, but only one smallish planet revolving around the sun, along with several others.

Earth is one of the nine planets—even that number is now in question; Pluto may actually belong to another solar system—that are held in the gravitational thrall of our life-giving star, the sun. Counting out from the sun, there are four *inner planets*. Rocky and smallish—on the cosmic scale, that is—they are also called the *terrestrial planets*.

Mercury Slightly larger than the earth's moon, Mercury is too small to have its own atmosphere. Without an atmosphere to trap heat, Mercury is very cold, even though it is the planet nearest the sun.

Venus The planet closest in size to Earth, Venus is shrouded by clouds that presumably create a greenhouse effect. The temperature on Venus is hot enough at 880°F (500°C) to melt lead. It is one of two planets on which spacecraft have landed (the Soviet Union having done it three times).

Earth (and its moon) The largest of the four terrestrial planets, Earth is the only one with liquid water on the surface and apparently the only one with tectonic activity.

Mars With about half the diameter of the earth, Mars has been called the Red Planet because the iron in its rocky surface has rusted, giving it a distinctive reddish patina. There are polar ice caps of dry ice (frozen carbon dioxide), and Mars also boasts the solar system's largest mountain, an extinct volcano called Olympus Mons that is 17 miles high and 370 miles across at its base. The notorious "canals" of Mars, first seen in 1877 and once believed to be a sign of life on the planet, simply don't exist and are most likely an optical illusion.

Between Mars and the other planets is a belt of interplanetary landfill, with celestial bodies ranging in size from tiny space pebbles to Ceres, an asteroid several hundred miles across. This is the asteroid belt formed of materials that under the right conditions might have formed another planet. Now it's a big mess of outer-space cookie crumbs, some of which occasionally cross paths with the earth.

Beyond the belt are the *outer planets*, called the *Jovian planets*, after the Roman name for Jupiter. Our knowledge of them has been transformed with the discoveries made by four unmanned space probes, the *Pioneers 10* and *11* and the *Voyagers 1* and *2*. We now know that these four large planets differ from the inner planets in that they may have a solid core but are surrounded by layers of liquid and gas.

Jupiter The largest planet spins rapidly—its "day" is only ten hours long. Scientists have speculated that since, like the sun, Jupiter is composed mainly of hydrogen and helium, and it has many moons—sixteen have been sighted so far—and is so large, it possibly was almost a star itself.

Saturn With its famous rings, Saturn is probably the most recognizable planet to laymen. The second-largest planet, it has twenty-one moons, one of which, Titan, is one of the largest satellites in the solar system and has its own atmosphere. Those seven notable ring systems are debris, mostly ice and rocks ranging in size from grains of dust to house-sized blocks.

Uranus Invisible to the naked eye, Uranus was not discovered until 1781 and was the first planet to be observed by telescope. The third-largest planet, it has fifteen moons, ten of them discovered by *Voyager 2*, and a series of dark rings.

Neptune First seen in 1846 and named for the Roman sea god, Neptune is the fourth-largest planet. With eight known moons and its own set of rings, Neptune also features the highest surface winds in the solar system, clocked at 1500 miles per hour. Its large moon, Triton, has an atmosphere and is the coldest known place in the solar system, measured at −390°F.

Pluto The ninth planet in our solar system is normally the planet farthest from the sun, usually about 3.7 billion miles. But an eccentricity in its orbit brought Pluto temporarily into a position inside Neptune's orbit during the 1990s. Along with its strange orbit, Pluto's small size (less than one-fifth the diameter of Earth) led some astronomers to contend that it is not a full-fledged planet. On August 24, 2006, the International Astronomical Union passed a new definition for planets and designated Pluto a "dwarf planet." Its one moon is called Charon.

Our solar system is centered, of course, on the sun, a whirling body of hydrogen and helium burning up in constant thermonuclear detonations. These explosions, which produce the light and heat that made life possible and sustain the earth today, start within the sun's center. The solar energy we now receive took thirty thousand years to work its way out from the solar core. Things get easier after that, though. Having reached the sun's surface, the solar energy takes but eight minutes to find its way across 93 million miles of empty space and down to earth. (It takes about five and a half hours for the sun's light to reach Pluto.)

Although it is our sun and we love it, the sun is a rather ordinary star. What seems to set it apart, of course, is the peculiar little planet called Earth that so far has proven to be the only place capable of generating and maintaining life.

The sun—and its tidy little solar system whirling about space with clockwork regularity—is but one of a large number of stars within a *galaxy*. Our galaxy is the Milky Way, which also explains the source of the word *galaxy*, derived from the Greek word *galaxias* for "milky." On a summer night, the reason for this name is apparent, as the Milky Way looks like a large, white puddle of milk, spilled across the sky.

Just as the sun is an average star, the Milky Way is in many ways a typical galaxy. It has about *100 billion* stars. Shaped somewhat like a child's pinwheel, the Milky Way spins through space with four spiral arms radiating out from the center. The closer to the center of the pinwheel, the more densely packed the stars. Our solar system lies within the Milky Way's Orion arm. It takes light a hundred thousand years to cross from one side of the Milky Way to the other.

Although humanity has known about the stars of the Milky Way for centuries, and philosopher Immanuel Kant speculated about the existence of other galaxies more than two hundred years ago, the discovery of galaxies was a recent one. Until 1923, in fact, other galaxies were thought to be clouds of gas. Edwin Hubble, one of the greatest of American astronomers, resolved the issue when he was able to discern individual stars in the Andromeda galaxy, the Milky Way's nearest galactic neighbor. Light from the Andromeda galaxy takes about two million years to reach our galaxy. Hubble also began a system of galaxy classification: the egg-shaped, or elliptical; the irregular; and the spiral, the form of the Milky Way.

Strewn across the empty darkness of space are perhaps a hundred billion galaxies, each containing millions or billions of stars. The galaxies, in turn, are collected in groups and superclusters. The Milky Way and nearby Andromeda belong to the friendly-sounding Local Group, which at twenty or thirty galaxies is fairly small potatoes in the intergalactic big picture. This Local Group lies at the fringes of the Virgo Supercluster, which consists of about five thousand galaxies. Lying between the superclusters are voids—vast areas of space where no stars shine. Unknown until the 1980s, the voids separating the superclusters of galaxies are like vast Saharas of space, except that they are millions of *light years* across.

So how big is the universe? To put it bluntly, it is an unimaginably big place. Astronomers can see out to distances of 10 billion light years, but that may be only part of the universe, the size of which is literally unknown. If you're still not impressed with space numbers, try and count the stars. The latest estimate for stars in the galaxies of the *visible* universe is 1 sextillion, or 1 followed by *21 zeroes*. And you thought the United States budget deficit was impressive. Wait till Washington, DC, hears that there is a number called sextillion.

How Far Is a Light Year?

Apart from our star, the sun, the star closest to earth is Proxima Centauri, a faint companion to the better-known Alpha Centauri. These stars are a mere 4.2 light years away, meaning it takes a little over four years for light from these stars to reach the earth. Or, to put it another way, that's more than *25 trillion* miles. That distance, by the way, is the average distance between most stars.

The distance between stars is so great that conventional measurements of the distance between them in familiar units is impractical. A more manageable astronomical unit, called the *light year*, was created to simplify matters. A light year is the distance it takes light to travel in one year at a speed of about 186,200 miles per second. One light year is equivalent to about 6 trillion miles.

Was There a Big Bang?

Most—but not all—astronomers will answer yes to this question. Like Darwin's "theories," the *Big Bang Theory*, which sets out to explain the creation of the universe, can only be disproven, not proven. But it is supported by a growing body of evidence. In science, a theory is not a good guess; it means a "well-supported explanation of some aspect of the natural world . . . repeatedly confirmed through observation and experiment." Nonetheless, it is always wise to remember that some of the best-educated people in the world believed the earth to be the center of the universe and thought Galileo was a nut-case and a heretic for his theories. It only takes a single discovery to shatter a world of preconceptions.

Basically, the Big Bang holds that time, space—all matter—began in one momentous instant: a hot, dense explosion that took place ten to twenty billion years ago. This was suggested in 1927 by a Belgian priest, Georges Lemaître, who proposed that the universe started with the explosion of a "primeval atom," in which all the mass of the universe had been concentrated in an extremely small space. In the first few *millionths of a second*, space expanded incredibly quickly for a very short time.

The notion of a big bang creation got a serious start from Edwin Hubble, the astronomer who discovered galaxies. At about the same time, Hubble determined that the galaxies were moving away from the earth. To be more precise, galaxies are moving away from each other. He also discovered that the farther away a galaxy is, the faster it is moving. In very simple terms, this was the basis for the theory that the universe is expanding.

Think of dropping a water balloon. It hits the ground and the water spreads outward from a central point of impact. The bigger the balloon, the more water—thus, the bigger the puddle that spreads out from the center. Like a cosmic water balloon sending out its immense puddle at the moment of impact, something had to propel the galaxies outward. That something was presumably the Big Bang.

The next major piece of evidence came in 1965 from two physicists working for the Bell Telephone Laboratories, Arno Penzias and Robert Wilson. While trying to refine their microwave-antenna equipment, the two men accidentally discovered a radio wave "hiss" that suggested to the scientists the remnants of the Big Bang.

Finally, in the spring of 1992, a team at the Lawrence Livermore Laboratory in California led by astrophysicist George Smoot announced a startling find that supported the Big Bang Theory. The new data came from Cosmic Background Explorer (COBE), a $160 million U.S. satellite launched in 1989 to measure microwave radiation in space. Describing their discovery as "wrinkles" in the smooth "fabric" of the universe, the team dated these wrinkles to 300,000 years after the Big Bang, a flash in cosmic time.

Now go back to the water balloon. Eventually, the water stops spreading. That is just one of many big, unanswered questions about the Big Bang and the future of the universe. Will the universe continue its expansion, moving outward infinitely? Or will it reach a limit, a point of maximum expansion after which the collective force of gravity will start pulling all those cosmological bodies yo-yo-ing back in toward the center? This possibility is called the *big crunch*. Whatever you think the answer is, you don't actually have to sweat it. We're looking at a time frame of several billion years here.

The new data also supports a theory that has been moving up and down the astrophysical popularity charts. It holds that much of the universe is composed of dark matter that we can't see.

Did an Asteroid Kill the Dinosaurs?

KILLER ASTEROIDS FROM SPACE! sounds like the title of a bad science fiction movie from the 1950s. Yet a planetary collision is not as outlandish as it may sound, even though some astronomers take it more seriously than others. And many dismiss the likelihood as so remote as not to warrant any serious expenditure of limited scientific resources or valuable research time. But it is an issue that the National Aeronautics and Space Administration (NASA) and Congress have taken seriously enough to study at some length. In congressional hearings in recent years, the results of those studies have been scary enough to make headlines and get editorial writers to scratching their heads.

First of all, look at precedent. We know that it happens. Although asteroids that make it into the atmosphere in the form of small meteors and meteorites usually either burn up or land harmlessly in the ocean or some unpopulated area. A falling stone occasionally punches a hole in someone's roof. But sometimes meteorites are not so small. A huge crater in Arizona was once presumed to have been caused by volcanic action, but it is now known to be the result of a meteor crashing to earth. There have been enough meteor crater discoveries elsewhere in the world to ascertain that it happens. A crater in southern Australia measures a hundred miles across. This didn't just happen millions of years ago, either. Once in this century—and possibly twice—the earth has had a "close encounter" with "incoming" from outer space.

The first of these recent events occurred in 1908 in the Tunguska region of Siberia. An explosion there leveled a huge area and destroyed millions of trees in a 1,200-square-mile forest. The trees left standing all bent away from a point of impact. But no traces of a fallen meteor were ever discovered—no physical evidence that anything actually hit the earth. The Tunguska event is now widely presumed to have been an asteriod or icy chunk of a comet glancing off the earth's atmosphere, at perhaps five miles above the earth's surface, and then skipping away into space, like a stone skimming across a lake. The destructive power of this cosmic fender-bender has been likened to the explosive force of twenty hydrogen bombs (without the radiation released by nuclear weapons). The explosion also lofted an enormous load of dust into the air and possibly destroyed a substantial portion of the protective ozone in the atmosphere. Scientists estimate the object that caused this destruction was about 150 feet in diameter.

The second impact is less certain. But in 1978, a huge explosion was detected in the South Atlantic. At the time, it was attributed to the detonation of an atomic bomb, perhaps a test by South Africa and/or Israel, both of which predictably denied such an event. Today, astronomers are guessing that what exploded was a rather small asteroid slamming into the earth with the equivalent of a hundred kilotons of TNT, far more destructive than the bomb that leveled Hiroshima.

Then, in 1991, astronomers using a telescope in Arizona called Spacewatch saw an asteroid pass silently by. At its nearest, the asteroid was about 106,000 miles away. That seems far, but not when you are talking about space and the intersection of planets. Had that asteroid, measuring about twenty-six feet across, hit the earth, it would have done so with the power of three Hiroshima bombs.

The grandest question of all is whether or not such an event had anything to do with the extinction of the dinosaurs. Many scientists now accept the likelihood that the extinction of the dinosaurs 65 million years ago was the result of the impact of an enormous asteroid crashing into the Caribbean basin near Yucatán in Mexico. This theory was first put forth by Walter and Luis Alvarez in the early 1980s. This killer asteroid kicked up a huge cloud of debris that plunged the entire world into darkness lasting months. The fireball produced a deadly chemical rain and would have altered the makeup of the atmosphere, perhaps for centuries—the actual time frame for the dinosaurs' demise.

This is the theory of *K–T extinction*—so called because the event supposedly took place at the time boundary between the Cretaceous (K in science shorthand) and Tertiary periods. It has been buttressed by the geological evidence of other extraterrestrial impacts that occurred in different time periods and led to similar mass extinctions.

Of course, this is theory and not everyone agrees. Many paleontologists accept the possibility of such an impact but refute the notion that it caused mass extinctions. One reason is that so many other animal forms alive at the same time were able to survive the conditions.

This is an interesting scientific parlor game that might be resolved before long. But it still leaves the question of whether it can—or might—happen again. To date, no person or animal has been killed by a meteorite—except for a dog hit by a meteorite in Egypt in 1911. If you are the gambling type, this question is all about playing the odds. The NASA team investigating the possibility found those odds slim but not negligible. They judged that the chances of an asteroid capable of inflicting serious global damage might hit the earth once in every 500,000 years. According to one estimate, that means for a person living seventy years, the chances of seeing the earth in upheaval as the result of an asteroid strike are 1 in 7,000. By comparison, the risk of death by automobile accident is much greater—1 in 100—but the risk of death by airplane crash is 1 in 20,000. As the old saying goes, "You pays your money and you takes your choice."

Milestones in Space Exploration

1919 American scientist Robert H. Goddard (1882–1945) publishes a paper, *A Method of Reaching Extreme Altitudes*, that suggests sending a small vehicle to the moon using rockets. When it is widely dismissed in the press, Goddard decides not to expose himself to further ridicule. His ideas are ignored by the U.S. government but attract great interest in Germany.

1926 Robert Goddard launches the first liquid-fuel-propelled rocket.

1929 American astronomer Edwin Hubble (1889–1953) establishes that the more distant a galaxy is, the faster it is receding from the earth. Known as Hubble's Law, this discovery confirms that the universe is expanding.

1930 The planet Pluto is discovered.

1938 Germany's liquid-fueled-rocket experiments, under the direction of engineer Wernher von Braun (1912–77), succeed in producing a rocket that can travel 11 miles (18 km). In 1944, improved models of these rockets are being used as the V-1 and V-2 self-propelled bombs launched against England in WWII. After the war, von Braun is brought to the United States and leads the development of the American rocket program that ultimately puts a man on the moon and conceives the space-shuttle concept.

1939 American physicist J. Robert Oppenheimer, who later headed the Manhattan Project that developed the atomic bomb during World War II, works out a theory of "black holes." Oppenheimer calculates that if the mass of a star is 3.2 times the mass of the sun, its collapse, caused by lack of internal radiation, would result in all the star's mass being reduced to a single point.

1949 A rocket-testing ground is set up at Cape Canaveral, Florida.

1957 The first space satellite, *Sputnik I*, is launched by the USSR in October. It is followed that year by *Sputnik II*, containing a live dog.

1961 The first man in space, Soviet cosmonaut Yuri Gagarin, is launched by the USSR and orbits the Earth.

1961 Alan B. Shepard becomes the first American in space, as his *Mercury 3* capsule completes a suborbital flight. Virgil I. Grissom becomes the second American in space.

1962 John Glenn is the first American to orbit the earth. Later that year, M. Scott Carpenter completes three orbits of the earth, and Walter M. Schirra completes six orbits.

1963 Soviet cosmonaut Valentina Tereshkova-Nikolayeva becomes the first woman in space.

1965 Arno Penzias and Robert Woodrow Wilson accidentally discover the radio-wave remnants of the Big Bang while trying to refine their radio equipment; their discovery convinces most astronomers that the Big Bang Theory is correct.

1966 The Soviet Union's *Venera III*, the first human-made object to land on another planet, reaches Venus.

1967 Three American astronauts—Virgil Grissom, Edward H. White II, and Roger Chaffee—die in a ground test of an Apollo spacecraft. That same year, Soviet cosmonaut Vladimir M. Komarov dies during the descent of his *Soyuz I* spacecraft.

1968 The first manned circumlunar flight by the *Apollo 8* mission makes ten orbits of the moon.

1969 *Mariner 6* passes close by Mars and returns television pictures and other data.

1969 On July 11, the American *Apollo 11* mission successfully lands on the moon. American astronaut Neil Armstrong becomes the first human being to stand on the moon. He was later joined by Colonel Edwin Aldrin. In November, *Apollo 12* makes the second manned lunar landing. A third landing is attempted in 1970 but is aborted because of equipment failure.

Geographic Voices   Neil Armstrong, first man to walk on the Moon

Of all the spectacular views we had, the most impressive to me was on the way to the Moon, when we flew through its shadow. We were still thousands of miles away, but close enough, so that the Moon almost filled our circular window. It was eclipsing the Sun, from our position, and the corona of the Sun was visible around the limb of the Moon as a gigantic lens-shaped or saucer-shaped light, stretching out to several lunar diameters. It was magnificent, but the Moon was even more so. We were in its shadow, so there was no part of it illuminated by the Sun. It was illuminated only by earthshine. It made the Moon appear blue-gray, and the entire scene looked decidedly three-dimensional. . . .

The sky is black, you know. It's a very dark sky. But it still seemed more like daylight than darkness as we looked out the window. It's a peculiar thing, but the surface looked very warm and inviting. It was the sort of situation in which you felt like going out there in nothing but a swimming suit to get a little sun. From the cockpit, the surface seemed to be tan. It's hard to account for that, because later when I held this material in my hand, it wasn't tan at all. It was black, gray and so on. It's some kind of lighting effect, but out the window the surface looks much more like light desert sand than black sand. . . .

1972 *Landsat I*, the first earth-resources satellite, is launched by the United States.

1972 The U.S. space probe *Pioneer 10* is launched. It will become the first human-made object to leave the solar system.

1973 *Pioneer 11* flies by Jupiter and Saturn, and discovers Saturn's eleventh moon and two new rings.

1973 Three U.S. Skylab missions are made.

1975 In June, Soviet *Venera 9*, an orbiter-lander, successfully orbits then lands on Venus, returning the first surface photo.

1975 In August, the U.S. *Viking 1* lands on Mars, the first American landing on another planet; it returns immense amounts of data. In September, *Viking 2* repeats this success. Its search for signs of life ends with ambiguous results.

1977 American space probes *Voyager 1* and *Voyager 2* are launched on a journey to Jupiter and the outer planets.

1980 A team led by Walter Alvarez and his father, Luis Walter Alvarez (1911–88), the 1968 Nobel Prize winner in physics, discovers a thin layer of clay at the Cretaceous–Tertiary (K–T) boundary (a time marked by mass extinction, including the dinosaurs). The clay is enriched with the heavy metal iridium, a compound common to meteors, leading the team to speculate that a giant body from space collided with the earth, causing the mass extinction, including that of dinosaurs. Later discoveries place the most likely point of impact in the Caribbean basin near Yucatán, Mexico.

1981 The first flight of the *Columbia*, the first space shuttle. The *Columbia* is relaunched later that year in the first reuse of a spacecraft.

1983 The second space shuttle, the *Challenger*, is successfully launched. On a second *Challenger* flight that year, Sally K. Ride becomes the first American woman in space. A third *Challenger* mission takes the first black astronaut, Guion S. Bluford Jr., into space.

1984 During the *Challenger's* fourth flight, two astronauts take the first untethered space walks using jet-propelled backpacks.

1986 The space shuttle *Challenger* blows apart seventy-three seconds after launch, killing six astronauts and civilian teacher Christa McAuliffe.

1989 NASA launches Mission to Planet Earth, an international scientific collaboration aimed at improving environmental data measurement.

1989 The U.S. *Magellan* maps Venus in unprecedented detail.

1990 *Voyager 1*, launched in 1977, flies by the Jupiter and Saturn systems and takes the first portrait of the solar system. The Hubble Space Telescope, a $2.1 billion orbiting telescope designed to see farther into space than ever before, is launched. Soon afterward, a flaw is discovered in one of its primary mirrors, seriously limiting the telescope's capabilities.

1992 Despite its flaws, the Hubble Telescope sends back information, including data on the hottest star ever recorded—thirty-three times hotter than our sun.

1992 A team at Lawrence Livermore Laboratory led by George Smoot announces the discovery of "ripples" in space that have been detected by the Cosmic Background Explorer (COBE), offering significant new evidence in support of the Big Bang Theory.

1992 Three astronauts working outside the newest space shuttle, the *Endeavour*, recapture a damaged satellite by hand in a daring and unprecedented space walk that demonstrates the value of the much-maligned shuttle program.

1995 *Galileo* becomes the first spacecraft to orbit Jupiter.

1996 *Mars Pathfinder* is launched and becomes the first spacecraft to land on Mars since 1976.

1998 The first component of the International Space Station (ISS) is launched.

1999 The world's most powerful X-ray telescope, the Chandra Observatory, is launched.

2000 *NEAR (Near Earth Asteroid Rendezvous) Shoemaker* becomes the first spacecraft to orbit and land on an asteroid (named Eros).

The first resident crew comes aboard the ISS.

2001 *Mars Odyssey*, an unmanned spacecraft, is launched by the United States; its findings include the presence of frozen water in the polar regions of Mars.

2003 *Mars Express* is launched. Its lander will examine rocks and solids on the surface of Mars.

The twin *Mars Exploration Rovers* are launched. These six-wheel robotic rovers find evidence of water in the Martian past.

2003 *Shenzou 5* becomes the first Chinese spacecraft to orbit Earth.

The space shuttle *Columbia* disintegrates while returning from a space flight, claiming the lives of seven crewmembers.

2004 *Spaceship One* becomes the first privately funded human spaceflight.

2005 *Deep Impact* becomes the first spacecraft to strike a comet's nucleus and study its interior composition.

2010 Launched by SpaceX, a commercial enterprise, the *Dragon* becomes the first private spacecraft to be launched into Earth's orbit and returned safely.

APPENDIX I
What the Hell Is a Hoosier?
Names and Nicknames of the Fifty American States

We see them all the time. State license plates with dopey mottoes and dumb nicknames. Did you ever find yourself wondering, "Who picked such a stupid nickname for their state?" This listing explains the derivation of the names of the states and attempts to sort out some of the silliness behind those occasionally absurd nicknames and state mottoes. The numbers following the names of the states refer either to their rank in age and date of ratification of the Constitution, for the first thirteen states, or rank and date of entrance into the Union for later states.

- *Alabama* (22nd state—1819) comes from a Choctaw Indian word meaning "thicket clearers" or "vegetation gatherers." Alabama is officially nicknamed the Yellowhammer State in honor of the state bird, the yellowhammer, better known as a woodpecker. During the Civil War, Alabama's soldiers also wore a yellow-tinged uniform, an odd color choice for a uniform.

- *Alaska* (49th state—1959) is a corruption of a native Aleut word meaning either "great land" or "that which the sea breaks against," and is now nicknamed the Last Frontier and Land of the Midnight Sun. U.S. secretary of state William Henry Seward arranged for the purchase of Alaska in 1867 from Russia for about two cents an acre. Alaska was first derided as "Seward's folly" or "Seward's icebox," but needless to say, the return on the investment has been substantial.

- *Arizona* (48th state—1912) derives from the Papago Indian word *arizonac*, meaning "little spring." An alternative suggestion is from the Spanish *arida zona*, for "dry land." Its official nickname, the Grand Canyon State, is a clear improvement over such other fond appellations as the State Where You Can Always Expect to Enjoy the Unexpected, the Valentine State (because it joined the Union on February 14, 1912), or the Italy of America, presumably coined for the state's scenic wonders. However, there are no Grand Canyons or Painted Deserts in Italy. Arizona is also the only state to boast of official neckwear—the bolo tie.

- *Arkansas* (25th state—1836) probably gets its name from *aken-zea*, a word of unknown meaning borrowed from the Quapaw, a Sioux tribe. Unfortunately, it has taken the rather dull Land of Opportunity as its official nickname instead of opting for more colorful alternatives such as the Guinea Pig State (the U.S. Department of Agriculture uses the state for its testing programs), the Place Where Plant Sites and Pine Forests Grow Side by Side, and the Toothpick State.

- *California* (31st state—1850) The origin of the Golden State's name is a bit obscure. The word was first used by Hernán Cortés in 1535 and may be derived from the Spanish words *caliente fornalla*, or "hot furnace." Another suggestion is that the name comes from the legendary isle of Greek myth ruled by Queen Caliphia. A third possibility is that the name is drawn from *Las Sergas de Esplandian*, a Spanish romance written in 1500, in which California is mentioned as an island near the Terrestrial Paradise.

- *Colorado* (38th state—1876) is derived from the Spanish word for "red" or "ruddy." Although officially nicknamed the Centennial State because it entered the Union in the year of America's Centennial, other popular names have included the Lead State (one of at least three Lead States), the Copper State, and the Switzerland of America.

- *Connecticut* (5th state—1788) is derived from another Indian name, Quinnehtukqut, meaning "beside the long tidal river." Its official designation as the Constitution State dates to the colonial period adoption of the *Fundamental Orders*, considered the first written American constitution, in 1639. The official nickname, the Nutmeg State, comes from the "wooden nutmeg," similar to a wooden nickel. State residents were supposed to be so smart they could sell wooden nutmegs to the gullible. Corporate giants who call the state home inspired the poetic appellation the State Where the Good Life Pays More in Corporate Dividends. Arms supplier to the nation since the time of the American Revolution, Connecticut has also been called the Arsenal of the Nation.

- *Delaware*, (1st state—1787) first of the thirteen original states to ratify the Constitution (1787), is the First State. The colony was initially named in 1610 for Virginia's governor, Thomas West, the Baron De La Warr. Also called the Small Wonder and the Diamond State (get it?—small, but valuable), the state's best nickname may be the Blue Hen State, in honor of a ferocious breed of fighting chicken popular during the revolutionary era.

- *Florida* (27th state—1845) was named by explorer Ponce de León in 1513 from the Spanish for "feast of flowers." For obvious reasons called the Sunshine State, Florida has been more appetizingly listed on the menu as the Winter Salad Bowl State.

- *Georgia* (4th state—1788), fourth state to ratify the Constitution (1788), was named in honor of King George II of England. Established as a colony in 1733 as a refuge for English debtors, its true purpose was to act as a defensive buffer between the Spanish outpost in Florida and the other existing English colonies to the north. Georgia's value was proven in 1742, when its founder, James Oglethorpe, led a successful defense against a Spanish invasion at the Battle of Bloody Marsh. The largest state east of the Mississippi, it is called the Peach State, even though other states grow more peaches, and the Goober State because it does grow the most peanuts. One leading peanut grower was President Jimmy Carter.

- *Hawaii* (50th state—1959) has uncertain origins. The collection of islands may have been named by Hawaii Loa, the person traditionally accepted as discoverer of the islands, or they may have been named after Hawaii, or Hawaiki, the home of the Polynesians who settled the islands sometime between the years 300 and 600. Known as the Aloha State, Hawaii is also called the Pineapple State for its most important agricultural product. A native monarchy was overthrown in 1893 and a republic was declared a year later with Sanford Dole as president. In 1898, Hawaii was formally annexed—a polite word for "taken over"— by the United States.

- *Idaho* (43rd state—1890) has the singular distinction of having a name that apparently doesn't mean anything. Although Idahi was a name given to the Comanche Indians by the Kiowa-Apaches, there is only speculation about its exact meaning. Two possibilities are "fish eaters" and "mountain gem." Another suggestion is that it is from the Shoshonean *Ee-dah-how*, which translates "Sunup!" or "Behold the Sun Coming Down the Mountain." Known as the Gem State for its many precious and semiprecious stones, it is also the Spud State because it produces—and processes—about one fourth of America's potato crop.

- *Illinois* (21st state—1818) comes from the Indian tribal name Inini, meaning "tribe of superior men" or "perfect and accomplished men" and adapted by the French as Illini. The Prairie State is also affectionately known as the Land of Lincoln because the sixteenth president, while born in Kentucky, eventually settled in Illinois and pursued his political career there.

- *Indiana* (19th state—1816) means simply "land of Indians," which is what it used to be. Although the state officially adopted the modest title "The Center of the Commercial Universe" in 1937, it is much better known as the Hoosier State, and its residents are called Hoosiers. The origin of the word is obscure, although some claim it comes from a common early pioneer greeting, "Who'shyer?" ("How are you?").

- *Iowa* (29th state—1846) is another obscure Indian word. It probably means "this is the place" or "the beautiful land," but others have suggested a meaning related to "cradle" or a derisive name for a tribe, meaning "sleepy." Schoolchildren should be happy that they settled on the name Iowa: a French map from 1763 spelled the name *Ouaouiatonon*. It is also called the Hawkeye State in honor of the Indian chief Hawkeye.

- *Kansas* (34th state—1861) comes from the Siouan word *kansa*, meaning "people of the south wind." Its unhappy name "Bleeding Kansas," or "Bloody Kansas," comes from the several years of vicious fighting between abolitionists and proslavery settlers who fought over the fate of the state as either a free or slave state. A most interesting nickname is the Salt of the Earth, as the state apparently has enough salt reserves to last for several hundred thousand years.

- *Kentucky* (15th state—1792) is derived from an Iroquoian word, *ken-tah-ten*, meaning "land of tomorrow." It is best known as the Bluegrass State for the native grass characteristic of its famed race-horse-breeding region.

- *Louisiana* (18th state—1812) is part of the large territory named by the French explorer La Salle for King Louis XIV, the Sun King. It is known as the Pelican State and the Sugar State.

- *Maine* (23rd state—1820) was thought to be first used to distinguish the "mainland" from the offshore islands but may also be associated with the English Queen Henrietta, wife of Charles I, who owned the French province of Mayne. With nearly 89 percent of its territory forested, it is easy to understand its nickname, the Pine Tree State.

- *Maryland* (7th state—1788) was named in honor of Henrietta Maria, wife of Charles II of England. Its only nicknames are the rather colorless monikers, the Free State and the Old Line State, but it does lay claim to an official state crustacean, the Maryland blue crab.

- *Massachusetts* (6th state—1788) comes from two Indian words meaning "great mountain place." The second British colony in America, it was founded as the Massachusetts Bay Colony, providing the obvious nicknames, the Bay State and the Old Colony State. Less appealing is its claim to being the Bean-Eating State.

- *Michigan* (26th state—1837) is also derived from two Indian words meaning "great lake." Michigan is also called the Wolverine State for the bushy-tailed animal that is not a wolf. In a bit of interstate one-upmanship, Michigan trumpets itself as the Wonderland of 11,000 Lakes, to top Minnesota (see below).

- *Minnesota* (32nd state—1858) gets its name from a Dakota Indian word meaning "sky-tinted water." The big question remains: is Minnesota truly the Land of 10,000 Lakes? And who counted them? If anything, 10,000 is a low figure. Estimates range from 11,000 to 15,000 lakes in the state also called the Gopher State and the Bread and Butter State.

- *Mississippi* (20th state—1817) is an Indian word meaning "father of the waters." It is officially called the Magnolia State.

- *Missouri* (24th state—1821) was named for the Missouri tribe, from the Indian word meaning "town of the large canoes." It is known as the Show-Me State, apparently reflecting the proudly skeptical character of its natives. For some reason, Missouri is also known as the Puke State. Wouldn't that make a nice license plate!

- *Montana* (41st state—1889) is derived from the Spanish for "mountainous." The fourth-largest state, it is officially nicknamed the Treasure State—mining played a significant part in its history and its official Latin motto is Oro y Plata, or Gold and Silver—but most folks prefer to think of it as the Big Sky State.

- *Nebraska* (37th state—1867) is from an Oto Indian word meaning "flat water." Best known as the Cornhusker State in honor of its most significant product and the farmers who harvest it, the state is also more colorfully known as the Bug-Eating State, in honor of bull bats, which eat bugs.

- *Nevada* (36th state—1864) comes from Spanish for "snow-capped." While its mountains have the snow, there isn't much rain. Nevada is the driest state, with desert areas that receive less than four inches of rain each year. Although known as the Silver State for the Comstock Lode, a rich silver deposit discovered in 1859, it is also called the Battle-Born State because it joined the Union during the Civil War.

- *New Hampshire* (9th state—1788) is named for the English county of Hampshire. Nicknamed the Granite State for the bedrock underlying its surface, it's probably better known for its defiant motto, *Live free or die.* It was the first colony to declare its independence from British rule.

- *New Jersey* (3rd state—1787) is named for the Isle of Jersey in the English Channel. Those only familiar with its urbanized and industrialized turnpike region are hard-pressed to understand its nickname, the Garden State.

- *New Mexico* (47th state—1912) is of course derived from the country of Mexico, from which it was severed after the Mexican War (1846–48). Cartoon fans will appreciate the fact that its official bird is the roadrunner ("Beep, Beep"). It also boasts an official state cookie, the bizcochito, and has been called the Vermin State, another of those nicknames that never made it onto a license plate.

- *New York* (11th state—1788) started its colonial history under Dutch rule as New Amsterdam, but was renamed in honor of the English Duke of York after the British took over the colony and threw out the Dutch governor. Officially the Empire State, lately people have been wondering if the empire will strike back.

- *North Carolina* (12th state—1789) was christened in honor of King Charles I of England. Its choice of nickname, the Tar Heel State, seems a little odd, since "tar heel" was an insulting name applied to the state's infantrymen during the Civil War by Mississippi soldiers who complained that the men from North Carolina didn't hold their positions—that is, they didn't put "tar on their heels."

- *North Dakota* (39th state—1889) is named after the Indian tribe called the Dakota, a word that means "allies." It is also known as the Flickertail State for an animal called the flickertail squirrel.

- *Ohio* (17th state—1803) comes from an Indian word meaning "great river." Best known as the Buckeye State for the horse chestnut, or buckeye, which resembles the eye of a buck. It has also been called the Modern Mother of Presidents, as it was the birthplace to seven of America's presidents: Ulysses S. Grant, Rutherford B. Hayes, James A. Garfield, Benjamin Harrison, William McKinley, William Howard Taft, and Warren G. Harding. (Virginia is the actual champion, having produced eight chief executives. See below.) The Ohioans in the White House were a fairly inconspicuous group, and unlucky—three of them died in office, two by assassination. New York and Massachusetts are runners-up with four apiece.

- *Oklahoma* (46th state—1907) comes from two Choctaw Indian words meaning "red people." Its nickname, the Sooner State, comes from those settlers who entered the territory "sooner" than the noon starting gun when the region was opened for homesteading on April 22, 1889.

- *Oregon* (33rd state—1859) has an uncertain derivation but was first used in 1778 by an English writer, Jonathan Carver. There are many suggested derivations, ranging from "hurricane" to "piece of dried apple." Another possibility is the Shoshonean word *oyer-un-gon*, which means "place of plenty." Nicknamed the Beaver State, Oregon was once also known as the Hard-Case State because life was so difficult for the early settlers, and the

Webfoot State because of the excessive rain in its Pacific coastal setting.

* *Pennsylvania* (2nd state—1787) was named by its founder, William Penn, in honor of his father, Admiral Sir William Penn, and means "Penn's woodland." Founded as a haven for Quakers who were being persecuted in England, it was long called the Quaker State. Geographically central as well as one of the most influential of the thirteen colonies, it is also called the Keystone State.

* *Rhode Island* (13th state—1790) was named in 1524 by the Italian explorer Verrazzano, who said it was about the size of the Greek island of Rhodes (see in Chapter 4, p. 168, "What Were the Seven Wonders of Antiquity?"). But later Dutch settlers also called it *rode* for "red," possibly referring to the color of the soil. The smallest state, it was founded by Roger Williams, exiled from the Massachusetts Bay Colony, on the basis of religious freedom, and was a haven for Quakers persecuted in England and the other Puritan colonies as well as Jews from Amsterdam (the Touro Synagogue, established in 1763, is the oldest in America).

* *South Carolina* (8th state—1788), like its sister to the north, is named for King Charles I of England. It is known as the Palmetto State after its official tree, the palmetto palm. It is the site of the first shots officially fired in the Civil War. South Carolina militiamen attacked Fort Sumter, a federal fort in the harbor at Charleston.

* *South Dakota* (40th state—1889) is named, like North Dakota, for the Dakota tribe (see above). In the Black Hills in the southwestern corner of the state is the famed Mount Rushmore in which the faces of Washington, Lincoln, Jefferson, and Theodore Roosevelt have been carved. New Mexico's roadrunner should be careful; South Dakota is the Coyote State.

- *Tennessee* (16th state—1796), is derived from a Cherokee word whose meaning is obscure. Once humbly dubbing itself the Nation's Most Interesting State, Tennessee has been more commonly known as the Volunteer State since 1847 when the governor called for 2,800 volunteers for the Mexican War and 30,000 men responded.

- *Texas* (28th state—1845) comes from an Indian word meaning "friends." The Texans were American southerners who had been invited into the territory by Mexico, which controlled it. But they rebelled because Mexico had emancipated slaves. Setting itself up as the Lone Star Republic, Texas fought for independence from Mexico in 1836, precipitating the Mexican War, after which the territory joined the Union.

- *Utah* (45th state—1896) is named after the Ute tribe, meaning "people of the mountains." First explored by Spanish priests, Utah was opened up to white settlers by the great migration of Mormons who came to escape religious persecution in the East beginning in 1847. The territory was part of the settlement made with Mexico in 1848.

- *Vermont* (14th state—1791) is a combination of the French words for "green mountain" which also provides its predictable nickname, the Green Mountain State, certainly not as interesting as the Land of Marble, Milk, and Honey.

- *Virginia* (10th state—1788), site of the first permanent English settlement in America—Jamestown, 1607—was the largest of the original thirteen colonies and certainly one of the most influential states in America's early history. Named in honor of Elizabeth I, the "virgin queen" of England, Virginia was also promoted by the men who wanted to attract new settlers to "virgin land." It was later called the Mother of Presidents, having produced four of the first five Presidents—George Washington, Thomas Jefferson, James Madison, and James Monroe—plus four others (William Henry Harrison, John Tyler, Zachary

Taylor, and Woodrow Wilson). Its political clout carried right up to the Civil War years, and it was the most significant of the Confederate states, being the site of the Confederate capital, Richmond, and the home state of Robert E. Lee, who led the Confederate troops.

- *Washington* (42nd state—1889) was named in honor of George Washington. It is also called the Chinook State after the tribe that inhabited the region. It is this tribe that also gives its name to a pair of separate winds: a characteristic warm, moist wind that blows from the sea on the Washington and Oregon coasts; as well as a warm, dry wind that descends on the eastern side of the Rocky Mountains and influences the climate of Colorado in particular.

- *West Virginia* (35th state—1863) originated as a part of Virginia. But after Virginia's secession from the Union in 1861, delegates from forty of the state's western counties formed their own government and were granted statehood as a non-slave state during the Civil War. Its rugged terrain includes the Appalachian and the Blue Ridge Mountains, source of its nickname, the Mountain State.

- *Wisconsin* (30th state—1848) is derived from a French corruption of an Indian word whose original meaning is lost. Also known as the Badger State for its official state animal, the burrowing badger, as well as for the fact that early settlers lived like badgers in sod homes that were partially underground.

- *Wyoming* (44th state—1890) comes from a Delaware Indian word meaning "mountains and valleys alternating," as does the Wyoming Valley in Pennsylvania. It is also known as the Equality State because it extended the vote to women in 1869.

APPENDIX II
Table of Comparative Measures

Miscellaneous Measurements

- *Acre*: Originally the area a yoke of oxen could plow in one day, it equals an area of 43,560 square feet (4,840 square yards).

- *Chain*: A chain is a surveying tool sometimes called a Gunter's, or surveyor's, chain. Measuring 66 feet long, it equals one tenth of a furlong and is divided into 100 parts called *links*. One mile is equal to 80 chains.

- *Cubit*: What was God talking about when he gave the ark's measurements to Noah? A cubit is 18 inches, or 45.72 cm. The length is derived from the distance between the elbow and the tip of middle finger and the word comes from the Latin *cubitum*, for elbow.

- *Knot*: A knot is not a distance but a measure of speed equal to 1 nautical mile per hour, used for measuring the speed of ships. The term is derived from the old sailor's practice of throwing a knotted rope over the side of the ship. The number of knots that were fed out in a specific period of time determined the ship's speed. The knotted rope was weighted at the end with a piece of wood, which is the source of the term "ship's log," in which the ship's speed, position, and other pertinent information were recorded.

- *League*: A rather indefinite and varying measure, but usually estimated at 3 statute miles in English-speaking countries. It comes from the medieval Latin word *leuga*, for "a measure of distance."

- *Mile*: In the United States and other English-speaking countries, a *statute* (or *land*) *mile* is a measure equal to 1,760 yards (5,280 feet). A *nautical mile* is a unit used for sea and air measurement and is equal to 1,852 meters or about 6,076 feet. The word originates in Latin *mille passuum*, for "a thousand paces."

- *Astronomical unit (A.U.)*: Used for astronomy because conventional measurements are unworkable, this is equal to 93 million miles, the average distance of the earth from the sun.

- *Light year*: A light year is 5,880,000,000,000 miles. Let's not be picky and just call it almost 6 trillion miles. This is the distance light travels in a vacuum in a year at the rate of 186,281.7 miles (299,792 kilometers) per second. If an astronomical unit (see above) were represented by one inch, a light year would be equivalent to about one mile. Used for measurements in interstellar space.

- *Parsec*: And you thought they made this up on *Star Trek*! The term is a combination of *par*allax and *sec*ond, and the distance is approximately 3.26 light years. It is used for measurement of interstellar distances.

Metric Measurements

Length

10 millimeters	=	1 centimeter
10 centimeters	=	1 decimeter
	=	100 millimeters
10 decimeters	=	1 meter
	=	1,000 millimeters
10 meters	=	1 decameter
10 decameters	=	1 hectometer
10 hectometers	=	1 kilometer
	=	1,000 meters

Area

100 square millimeters	=	1 square centimeter
10,000 square centimeters	=	1 square meter
100 square meters	=	1 are
100 ares	=	1 hectare
	=	10,000 square meters
100 hectares	=	1 square kilometer
	=	1,000,000 square meters

Customary U.S. Measures

Length

12 inches	=	1 foot
3 feet	=	1 yard
5-½ yards	=	1 rod
40 rods	=	1 furlong

	=	220 yards (660 feet)
8 furlongs	=	1 land (statute) mile
	=	1,760 yards (5,280 feet)
3 land (statute) miles	=	1 league
1 international nautical mile	=	6,076.11549 feet

Area

144 square inches	=	1 square foot
9 square feet	=	1 square yard
30-¼ square yards	=	1 square rod
	=	272 ¼ square feet
160 square rods	=	1 acre
	=	4,840 square yards (43,560 square feet)
640 acres	=	1 square mile

Surveyor's Chain Measure

7.92 inches	=	1 link
100 links	=	1 chain
	=	4 rods
	=	66 feet
80 chains	=	1 statute mile
	=	320 rods
	=	5,280 feet

Metric and U.S. Equivalents Length

1 centimeter	=	0.3937 inch
1 decimeter	=	3.937 inches
1 decameter	=	32.808 feet
1 fathom	=	6 feet
	=	1.8288 meters
1 cable's length	=	120 fathoms

	=	720 feet
	=	219.456 meters
1 foot	=	0.3048 meters
1 furlong	=	10 chains
	=	660 feet
	=	220 yards
	=	⅛ statute mile
	=	201.168 meters
1 inch	=	2.54 centimeters
1 kilometer	=	0.621 mile
1 league (land)	=	3 statute miles
	=	4.828 kilometer
1 chain	=	66 feet
	=	20.1168 meters
1 link	=	7.92 inches
1 meter	=	39.37 inches
	=	1.094 yards
1 mile (statute, or land)	=	5,280 feet
	=	1.609 kilometers
1 nautical mile (international)	=	1.852 kilometers
	=	1.151 statute miles
1 millimeter	=	0.03937 inch
1 yard	=	0.9144 meters

Area or Surface

1 acre	=	43,560 square feet
	=	4,840 square yards
	=	0.405 hectare
1 hectare	=	2.471 acres
1 square centimeter	=	0.155 square inch
1 square foot	=	929.030 square centimeters
1 square inch	=	6.4516 square centimeters
1 square kilometer	=	0.386 square mile

	=	247.105 acres
1 square meter	=	1.196 square yards
	=	10.764 square feet
1 square mile	=	258.999 hectares
1 square yard	=	0.836 square meters

APPENDIX III
The Nations of the World

The following list gives the names of the 193 member states (and one observer state) of the United Nations, divided by continent. See www.un.org/en/members/index.shtml.

A good source of basic geographic and economic information is the World Factbook on the Central Intelligence Agency's website: www.cia.gov/library/publications/the-world-factbook/geos/ek.html.

Africa

Algeria
Angola
Benin
Botswana
Burkina Faso
Burundi
Cameroon
Cape Verde
Central African Republic
Chad
Comoros
Congo
Côte d'lvoire (Ivory Coast)
Djibouti
Egypt

Equatorial Guinea
Eritrea
Ethiopia
Gabon
Gambia
Ghana
Guinea
Guinea-Bissau
Kenya
Lesotho
Liberia
Libya
Madagascar
Malawi
Mali
Mauritania
Mauritius
Morocco
Mozambique
Namibia
Niger
Nigeria
Rwanda
São Tomé and Príncipe
Senegal
Seychelles
Sierra Leone
Somalia
South Africa
South Sudan (gained independence from Sudan in July 2011)
Sudan
Swaziland
Tanzania
Togo
Tunisia
Uganda
Zaire

Zambia
Zimbabwe

————

Asia

Afghanistan
Azerbaijan (former Soviet republic)
Bahrain
Bangladesh
Bhutan
Brunei Darussalam
Cambodia
China
Cyprus
Democratic People's Republic of Korea (North Korea: claimed by
    South Korea)
East Timor
India
Indonesia
Iran
Iraq
Israel
Japan
Jordan
Kazakhstan (formerly a Soviet republic)
Kuwait
Kyrgyzstan (formerly a Soviet republic)
Laos
Lebanon
Malaysia
Maldives
Mongolia
Myanmar
Nepal

Oman
Pakistan
Palau
Philippines
Qatar
Republic of Korea (South Korea: claimed by North Korea)
Saudi Arabia
Singapore
Sri Lanka
Syria
Tajikistan (formerly a Soviet Republic)
Thailand
Turkey
Turkmenistan (formerly a Soviet Republic)
United Arab Emirates
Uzbekistan (formerly a Soviet Republic)
Vietnam
Yemen

Palestine is a declared state and has received diplomatic recognition from more than 100 states. It is a member of one specialized United Nations agency (UNESCO) but is disputed by Israel. It is administered by the Palestinian National Authority and represented by the Palestine Liberation Organization, which has permanent observer status at the UN.

Taiwan, the Republic of China, is a former UN state claimed by the People's Republic of China and participates in some United Nations-related organizations.

Australia (including Oceania, or the Pacific Islands)

Australia
Federated States of Micronesia (comprised of the islands of Yap, Chuuk, Pohnpel, and Kosrae)

Fiji
Kiribati
Marshall Islands
Nauru
New Zealand
Papua New Guinea
Samoa
Solomon Islands
Tonga
Tuvalu

Europe

Albania
Andorra
Armenia (considered part of "Eurasia," Armenia has applied to the
    European Union and is listed here accordingly)
Austria
Belarus (formerly Byelorussia, a Soviet Republic)
Belgium
Bosnia and Herzegovina (formerly a Yugoslavian republic)
Bulgaria
Croatia (formerly a Yugoslavian republic)
Czech Republic
Denmark
Estonia
Finland
France
Georgia (formerly a Soviet republic also considered part of "Eur-
    asia," it is a member of the Council of Europe and listed here
    accordingly)
Germany
Greece
Hungary

Iceland
Ireland
Italy
Latvia
Liechtenstein
Lithuania
Luxembourg
Macedonia (a republic that was formerly part of Yugoslavia)
Malta
Moldova
Monaco
Montenegro
Netherlands
Norway
Poland
Portugal
Romania
Russia (lies in both Europe and Asia)
San Marino
Serbia
Slovakia (formerly part of Czechoslovakia)
Slovenia (formerly a Yugoslavian republic)
Spain
Sweden
Switzerland
Ukraine (formerly a Soviet republic)
United Kingdom

Kosovo, a republic that declared its independence from Serbia in 2008, is recognized by the UN; Serbia maintains its sovereignty.

The Vatican City State is a United Nations observer state.

North America (including the Caribbean Islands)

Antigua and Barbuda
Bahamas
Barbados
Canada
Costa Rica
Cuba
Dominica
Dominican Republic
Grenada
Haiti
Jamaica
Mexico
St. Kitts and Nevis
St. Lucia
St. Vincent and the Grenadines
Trinidad and Tobago
United States of America

South America (including Central America)

Argentina
Belize
Bolivia
Brazil
Chile
Colombia
Costa Rica
Ecuador
El Salvador
Guatemala

Guyana
Honduras
Nicaragua
Panama
Paraguay
Peru
Suriname
Uruguay
Venezuela

Acknowledgments

I would like to thank many people who made this book possible from inception to finished pages, and now to a revised edition. My original editor, Mark Gompertz, who has been there from the beginning, deserves a large measure of gratitude. I also wish to thank Adrian Zackheim, Suzanne Oaks, Lisa Considine, Sharyn Rosenblum, and all the other people at William Morrow and Avon Books who have been so helpful. At HarperCollins, I have been lucky to work with Michael Signorelli, Jen Hart, Erica Barmash, Diane Burrowes, and Virginia Stanley, among many others. Thank you to copyeditor Ted Gilley for your excellent work.

For his friendship and support, I thank my good friend David Black, who is also my literary agent, but that comes after friendship.

I am very grateful to Edward Bergman of Lehman College, who provided so much help in my original edition of this book. If the rest of the world's geography teachers were as entertaining and interesting as he is, this book might be unnecessary.

This book would also not be possible without the resources of the New York Public Library as well as the library and archives at the American Museum of Natural History.

I'd like to also thank my understanding and supportive friends, in particular the late Margaret Enoch, whose encouragement has always meant a great deal to me.

My greatest gratitude goes to my wife, Joann, for her wisdom, creativity, patience, and support. She has made it all possible. And thank you, Jenny and Colin, for being so patient with a daddy who is at his desk too much of the time. (And Jenny, thanks for the sea horses!)

Bibliography

The following list of books includes a list of general readings and specific sources for each chapter. I have attempted to limit this list to books that are readily available and which represent the most up-to-date information available on the subjects discussed.

The list includes a section of books on geographic subjects that are suitable for children.

General Reference

al Faruqi, Isma'il Ragi A., ed. *Historical Atlas of the Religions of the World.* New York: Macmillan, 1974.

Blandford, Percy W. *Maps and Compasses: A User's Handbook.* Blue Ridge Summit, Penn.: Tab Books, 1984.

Carey, John. *Eyewitness to History.* Cambridge, MA: Harvard University Press, 1987.

Comrie, Bernard, Stephan Matthews, and Maria Polinsky, editors. *The Atlas of Languages: The Origin and Development of Languages Throughout the World* (rev. ed.). New York: Facts on File, 2003.

Davis, Kenneth C. *Don't Know Much About® History: Everything You Need to Know About American History but Never Learned* (rev. ed.). New York: Harper, 2011.

———. *Don't Know Much About the Bible: Everything You Need to Know About the Good Book but Never Learned.* New York: Morrow, 1998.

———. *Don't Know Much About® Mythology.* New York: HarperCollins, 2005.

———. *Don't Know Much About® the Universe.* New York: HarperCollins, 2001.

De Blij, Harm J., and Peter O. Muller. *Geography: Regions and Concepts* (5th ed.). New York: Wiley, 1988.

De Blij, Harm. *Why Geography Matters: Three Challenges Facing America: Climate Change, the Rise of China, and Global Terrorism.* New York: Oxford University Press, 2005.

Diamond, Jared. *Guns, Germs, and Steel: The Fates of Human Societies.* New York: Norton, 1998.

———. *Collapse: How Societies Choose to Fail or Succeed.* New York: Viking, 2005.

Espenshade, Edward B. Jr., ed. *Goode's World Atlas* (18th ed.). Chicago: Rand McNally, 1990.

Goodall, Brian. *The Penguin Dictionary of Human Geography.* London: Penguin, 1987.

Gould, Peter, and Rodney White. *Mental Maps.* Baltimore: Penguin, 1974.

Grigson, Lionel. *Wonders of the World.* New York: Gallery Books, 1985.

Grillet, Donnat V. *Where on Earth: A Refreshing View of Geography.* New York: Prentice Hall, 1991.

Hazen, Robert M., and James Trefil. *Science Matters: Achieving Scientific Literacy.* Garden City, NY: Doubleday, 1991.

Hellemans, Alexander, and Bryan Bunch. *The Timetables of Science: A Chronology of the Most Important People and Events in the History of Science.* New York: Simon & Schuster, 1988.

Kapit, Wynn. *The Geography Coloring Book.* New York: Harper-Collins, 1991.

Kidron, Michael, and Ronald Segal. *The New State of the World Atlas.* New York: Simon & Schuster, 1987.

Kjellstrom, Bjorn. *Be Expert with Map and Compass: The Complete "Orienteering" Handbook.* New York: Scribners, 1976.

Lacey, Peter, ed. *Great Adventures That Changed the World: The World's Great Explorers, Their Triumphs and Tragedies.* Pleasantville, NY: Reader's Digest Association, 1978.

McKnight, Tom L. *Physical Geography: A Landscape Appreciation* (3rd ed.). Englewood Cliffs, NJ: Prentice-Hall, 1990.

Makower, Joel, ed. *The Map Catalog* (2nd ed.). New York: Vintage, 1990.

Manguel, Alberto, and Gianni Guadalupi. *The Dictionary of Imaginary Places.* New York: Macmillan, 1980.

Marshall, Bruce, ed. *The Real World: Understanding the Modern World Through the New Geography.* Boston: Houghton Mifflin, 1991.

Milne, Courtney. *The Sacred Earth.* Saskatoon, Saskatchewan: Prairie Books, 1991.

Monmonier, Mark. *How to Lie with Maps.* Chicago: University of Chicago Press, 1991.

Moore, W. G. *The Penguin Dictionary of Geography* (7th ed.). New York: Penguin, 1988.

Morrison, Philip, and Phylis Morrison, *The Ring of Truth: An Inquiry into How We Know What We Know.* New York: Vintage, 1989.

Muller, Robert, and Theodore M. Oberlander. *Physical Geography Today: A Portrait of a Planet* (2nd ed.). New York: Random House, 1978.

Munro, David, ed. *Cambridge World Gazetteer*. New York: Cambridge University Press, 1988.

Rand McNally Desk Reference World Atlas. Chicago: Rand McNally, 1987.

National Assessment Governing Board: U.S. Department of Education. *Geography Framework for the 2010 National Assessment of Education Progress*. Washington, DC: National Assessment Governing Board, 2010.

Roberts, David. *Great Exploration Hoaxes*. San Francisco: Sierra Club Books, 1991.

Room, Adrian. *Place Names of the World*. London: Angus & Robertson, 1987.

Smart, Ninian. *World Religions* (2nd ed.). New York: Cambridge University Press, 1998.

Smith, Huston. *The Illustrated World's Religions: A Guide to Our Wisdom Traditions*. New York: HarperCollins, 1994.

Stewart, George R. *Names of the Land: A Historical Account of Place-Naming in the United States*. Boston: Houghton Mifflin, 1958.

Stoddart, D. R. *On Geography and Its History*. New York: Basil Blackwell, 1986.

Trefil, James. *1001 Things Everyone Should Know About Science*. New York: Doubleday, 1992.

Westwood, Jennifer. *The Atlas of Mysterious Places*. New York: Weidenfeld & Nicolson, 1987.

Chapter One

Bellonci, Maria. Translated by Teresa Waugh. *The Travels of Marco Polo*. New York: Facts on File, 1984.

Bergreen, Laurence. *Columbus: The Four Voyages*. New York: Viking, 2011.

Berthon, Simon, and Andrew Robinson. *The Shape of the World: The Mapping and Discovery of the Earth*. Chicago: Rand McNally, 1991.

Boorstin, Daniel J. *The Discoverers*. New York: Random House, 1983.

———. *The Seekers: The Story of Man's Continuing Quest to Understand His World*. New York: Random House, 1998.

Brown, Lloyd A. *The Story of Maps*. Boston: Little, Brown, 1949.

Campbell, Tony. *The Earliest Printed Maps: 1472–1500*. Berkeley: University of California Press, 1987.

Cohen, J. M., ed. *The Four Voyages of Christopher Columbus*. New York: Penguin, 1969.

Galanopoulos, A. G., and Edward Bacon. *Atlantis: The Truth Behind the Legend.* New York: Bobbs-Merrill, 1969.

Granzotto, Gianni. *Christopher Columbus.* Garden City, NY: Doubleday, 1985.

Hale, John R., et al. *Age of Exploration.* New York: Time Inc., 1966.

James, Preston E., and Geoffrey J. Martin. *All Possible Worlds: A History of Geographical Ideas.* New York: Wiley, 1972.

Litvinoff, Barnet. *1492: The Decline of Medievalism and the Rise of the Modern Age.* New York: Scribners, 1991.

Marshall, P. J., and Glyndwr Williams. *The Great Map of Mankind: Perceptions of New Worlds in the Age of Enlightenment.* Cambridge, MA: Harvard University Press, 1982.

Mavor, James W. Jr. *Voyage to Atlantis: A Firsthand Account of the Scientific Expedition to Solve the Riddle of the Ages.* Rochester, VT.: Park Street Press, 1990.

Morison, Samuel Eliot. *Admiral of the Ocean Sea: A Life of Christopher Columbus.* Boston: Little, Brown, 1942.

Newby, Eric. *The Rand McNally World Atlas of Exploration.* London: Mitchell Beazley Publishers, 1975.

Pellegrino, Charles. *Unearthing Atlantis: An Archeological Odyssey.* New York: Random House, 1991.

Ptolemy, Claudius. *The Geography.* New York: Dover, 1991.

Sale, Kirkpatrick. *The Conquest of Paradise: Christopher Columbus and the Columbian Legacy.* New York: Knopf, 1990.

Sobel, Dava. *Longitude: The True Story of a Scientific Genius Who Solved the Greatest Scientific Problem of His Time.* New York: Walker, 1995.

Viola, Herman J., and Carolyn Margolis, eds. *Seeds of Change: A Quincentennial Commemoration.* Washington, DC: Smithsonian Institution Press, 1991.

Wilford, John Noble. *The Mapmakers: The Story of the Great Pioneers in Cartography from Antiquity to the Space Age.* New York: Knopf, 1981.

———. *The Mysterious History of Christopher Columbus: An Exploration of the Man, the Myth, the Legacy.* New York: Knopf, 1991.

Chapter Two

Aveni, Anthony. *Empires of Time: Calendars, Clocks, and Cultures.* New York: Basic Books, 1989.

Erickson, Jon. *Volcanoes and Earthquakes.* Blue Ridge Summit, PA: Tab Books, 1988.

Flaste, Richard, ed. *The New York Times Book of Scientific Literacy*. New York: Times Books, 1991.

Gould, Stephen Jay. *Time's Arrow, Time's Cycle: Myth and Metaphor in the Discovery of Geological Time*. Cambridge, MA: Harvard University Press, 1987.

———. *Wonderful Life: The Burgess Shale and the Nature of History*. New York: Norton, 1989.

Hartmann, William K. *The History of the Earth: An Illustrated Chronicle of an Evolving Planet*. New York: Workman Publishing, 1991.

Hughes, Robert. *The Fatal Shore: The Epic of Australia's Founding*. New York: Knopf, 1986.

Krishtalka, Leonard. *Dinosaur Plots and Other Intrigues in Natural History*. New York: Morrow, 1989.

Lavender, David. *The Way to the Western Sea: Lewis and Clark Across the Continent*. New York: Harper & Row, 1988.

Lindsay, William. *The Great Dinosaur Atlas*. Englewood Cliffs, NJ: Julian Messner, 1991.

McPhee, John. *Basin and Range*. New York: Farrar, Straus, 1980.

———. *Rising from the Plains*. New York: Farrar, Straus, 1986.

Moorehead, Alan. *The Fatal Impact: An Account of the Invasion of the South Pacific*. New York: Harper & Row, 1966.

Revkin, Andrew. *The Burning Season: The Murder of Chico Mendes and the Fight for the Amazon Rain Forest*. Boston: Houghton Mifflin, 1990.

Røhr, Anders, ed. *Earth History*. Denver, CO: Earthbooks, 1991.

Shoumatoff, Alex. *The Rivers Amazon*. San Francisco: Sierra Club Books, 1978, 1986.

———. *The World Is Burning: Murder in the Rain Forest*. Boston: Little, Brown, 1990.

Snyder, Gerald S. *In the Footsteps of Lewis and Clark*. Washington, DC: National Geographic Society, 1970.

Weiner, Jonathan. *Planet Earth*. New York: Bantam Books, 1986.

Young, Louise B. *The Blue Planet: A Celebration of the Earth*. Boston: Little, Brown, 1983

Chapter Three

Bascom, Willard. *The Crest of the Wave: Adventures in Oceanography*. New York: Harper & Row, 1988.

Carson, Rachel. *The Sea Around Us.* New York: Oxford University Press, 1951.

Elder, Danny, and John Pernetta, eds. *The Random House Atlas of the Oceans.* New York: Random House, 1991.

Erickson, Jon. *The Mysterious Oceans.* Blue Ridge Summit, PA: Tab Books, 1988.

Groves, Don. *The Oceans: A Book of Questions.* New York: Wiley, 1989.

Kurlansky, Mark. *Cod: A Biography of the Fish That Changed the World.* New York: Walker Books, 1997.

McLynn, Frank. *Captain Cook: Master of the Seas.* New Haven: Yale University Press, 2011.

Winchester, Simon. *Atlantic: Great Sea Battles, Heroic Discoveries, Titanic Storms, and a Vast Ocean of a Million Stories.* New York: Harper, 2010.

Chapter Four

Acemoglu, Daron, and James A Robinson. *Why Nations Fail: The Origins of Power, Prosperity, and Poverty.* New York: Crown, 2012.

Adams, Mark.   *Turn Right at Machu Picchu: Rediscovering the Lost City One Step at a Time.* New York: Dutton, 2011.

Bierhorst, John. *The Mythology of North America.* New York: Morrow, 1985.

———. *The Mythology of South America.* New York: Morrow, 1988.

———. *The Mythology of Mexico and Central America.* New York: Morrow, 1991.

Blacker, Irwin R., ed. *The Portable Hakluyt's Voyages.* New York: Viking, 1965.

Brown, Michael H. *The Search for Eve.* New York: Harper & Row, 1990.

Campbell, Joseph. *Historic Atlas of World Mythology, Volume I: The Way of the Animal Powers, Part 1: The Mythologies of the Primitive Hunters and Gatherers.* New York: Harper & Row, 1988.

———. *Historic Atlas of World Mythology, Volume I: The Way of the Animal Powers, Part 2: Mythologies of the Great Hunt.* New York: Harper & Row, 1988.

———. *Historical Atlas of World Mythology, Volume II: The Way of the Seeded Earth, Part 1: The Sacrifice.* New York: Harper & Row, 1988.

———. *Historical Atlas of World Mythology, Volume II: The Way of the Seeded Earth, Part 2: Mythologies of the Primitive Planters: The Northern Americas.* New York: Harper & Row, 1989.

———. *Historical Atlas of World Mythology, Volume II: The Way of the Seeded Earth, Part 3: Mythologies of the Primitive Planters: The Middle and Southern Americas.* New York: Harper & Row, 1989.

Darwin, Charles. *The Voyage of the Beagle*, with an Introduction by Walter Sullivan. New York: Mentor Books/New American Library, 1972.

Dolnick, Edward. *Down the Great Unknown: John Wesley Powell's 1869 Journey of Discovery and Tragedy Through the Grand Canyon*. New York: HarperCollins, 2001.

East, Gordon W. *The Geography Behind History*. New York: Norton, 1967.

Freeman, Charles. *Egypt, Greece and Rome: Civilizations of the Ancient Mediterranean* (2nd ed.). New York: Oxford University Press, 2004.

Friedman, Thomas L. *Hot, Flat, and Crowded: Why We Need a Green Revolution and How It Can Renew America*. New York: Farrar, Straus & Giroux, 2008.

———. *The Lexus and the Olive Tree: Understanding Globalization*. New York: Anchor Books, 2000.

Friedman, Thomas L., and Michael Mandelbaum. *That Used to Be Us: How America Fell Behind in the World It Invented and How We Can Come Back*. New York: Farrar, Straus & Giroux, 2011.

Fromkin, David. *A Peace to End All Peace: The Fall of the Ottoman Empire and the Creation of the Modern Middle East*. New York: Holt, 1989.

Garreau, Joel. *The Nine Nations of North America*. Boston: Houghton Mifflin, 1981.

Grann, David. *The Lost City of Z: A Tale of Deadly Obsession in the Amazon*. New York: Doubleday, 2009.

Harris, Marvin. *Our Kind: Who We Are, Where We Came From and Where We Are Going*. New York: Harper & Row, 1989.

Humble, Richard. *Famous Land Battles: From Agincourt to the Six-Day War*. Boston: Little, Brown, 1979.

Ingpen, Robert, and Philip Wilkinson. *Encyclopedia of Events That Changed the World: 80 Turning Points in History*. New York: Viking Studio Books, 1991.

Johanson, Donald, and James Shreeve. *Lucy's Child*. New York: Morrow, 1989.

Keegan, John. *The Face of Battle*. New York: Viking, 1976.

Kennedy, Paul. *The Rise and Fall of Great Powers: Economic Change and Military Conflict from 1500 to 2000*. New York: Random House, 1987.

Kinzler, Stephen. *Overthrow: America's Century of Regime Change from Hawaii to Iraq*. New York: Times Books, 2006.

Klare, Michael T. *The Race for What's Left: The Global Scramble for the World's Last Resources*. New York: Metropolitan/Holt, 2012.

———. *Resource Wars: The New Landscape of Global Conflict*. New York: Henry Holt, 2001.

————. *Rising Powers, Shrinking Planet: The New Geopolitics of Energy.* New York: Metropolitan/Holt, 2008.

Kotkin, Joel. *The Next Hundred Million: American in 2050.* New York: Penguin, 2010.

Kurlansky, Mark. *Salt: A World History.* New York: Walker Books, 2002.

Lasky, Kathryn. Illustrated by Whitney Powell. *Traces of Life: The Origins of Humankind.* New York: Morrow Junior Books, 1989.

Lewin, Roger. *Bones of Contention: Controversies in the Search for Human Origins.* New York: Simon & Schuster, 1987.

Macdonald, John. *Great Battlefields of the World.* New York: Collier Books, 1984.

McEvedy, Colin. *The Penguin Atlas of Ancient History.* London: Penguin Books, 1967.

————. *The Penguin Atlas of Medieval History.* London: Penguin, 1961.

————. *The Penguin Atlas of Modern History to 1815.* London: Penguin Books, 1972.

————. *The Penguin Atlas of North American History to 1870.* London: Penguin Books, 1988.

MacGregor, Neil. *A History of the World in 100 Objects.* New York: Viking, 2010.

Mann, Charles. *1491: New Revelations of the Americas Before Columbus.* New York: Knopf, 2005.

————. *1493: Uncovering the New World Columbus Created.* New York: Knopf, 2011.

Menzies, Gavin. *1421: The Year China Discovered America.* New York: Morrow, 2003.

————. *1434: The Year a Magnificent Chinese Fleet Sailed to Italy and Ignited the Renaissance.* New York: Morrow, 2008.

Millard, Candice. *The River of Doubt: Theodore Roosevelt's Darkest Journey.* New York: Doubelday, 2005.

Miller, Jonathan. Illustrated by Borin Van Loon. *Darwin for Beginners.* New York: Pantheon, 1982.

Mithen, Steven. *After the Ice: A Global Human History, 20,000–5000 BC.* Cambridge, MA: Harvard University Press, 2004.

Núñez Cabeza de Vaca, Alvar, edited by Enrique Pupo-Walker, translated by Frances M. López-Morillas. *Castaways: The Narrative of Alvar Núñez Cabeza De Vaca.* Berkeley: University of California Press, 1993.

Oliver, Roland, and J. D. Farge. *A Short History of Africa* (6th ed.) New York: Penguin, 1988.

Pakenham, Thomas. *The Scramble for Africa: The White Man's Conquest of the Dark Continent from 1876 to 1912*. New York: Random House, 1991.

Parkman, Francis. *The Oregon Trail*. New York: New American Library, 1950.

Pauketat, Timothy R. *Cahokia: Ancient America's Great City on the Mississippi*. New York: Viking, 2009.

Pritchard, James B., ed. *The Harper Concise Atlas of the Bible*. New York: HarperCollins, 1991.

Raup, David M. *Extinction: Bad Genes or Bad Luck?* New York: Norton, 1991.

Reader's Digest Association. *The Last Two Million Years*. Pleasantville, NY: Reader's Digest Association, 1973.

Reston, James Jr. *Dogs of God: Columbus, the Inquisition, and the Defeat of the Moors*. New York: HarperCollins, 2005.

Sandlin, Lee. *Wicked River: The Mississippi When It Last Ran Wild*. New York: Pantheon, 2010.

Shorto, Russell. *The Island at the Center of the World: The Epic Story of Dutch Manhattan and the Forgotten Colony That Shaped America*. New York; Doubleday, 2004.

Steffen, Alex, ed. *Worldchanging: A User's Guide for the 21st Century*. New York: Abrams, 2011.

Stein, Mark. *How the States Got Their Shapes*. New York: Smithsonian/HarperCollins, 2008.

———. *The People Behind the Border-Lines: How the States Got Their Shapes Too*. Washington, DC: Smithsonian Books, 2011.

Taylor, Peter J. *Political Geography: World-Economy, Nation-State, and Locality*. London: Longman, 1985.

Tuchman, Barbara W. *The March of Folly: From Troy to Vietnam*. New York: Knopf, 1984.

White, Richard. *Railroaded: The Transcontinentals and the Making of Modern America*. New York: Norton, 2011.

Wilson, Ian. *Before the Flood: The Biblical Flood As a Real Event and How It Changed the Course of Civilization*. New York: St. Martin's, 2001.

Winchester, Simon. *A Crack in the Edge of the World: America and the Great Earthquake of 1906*. New York: HarperCollins, 2005.

———. *Krakatoa: The Day That the World Exploded: August 27, 1883*. New York: HarperCollins, 2003.

———. *The Man Who Loved China: The Fantastic Story of the Eccentric Scientist Who Unlocked the Mysteries of the Middle Kingdom*. New York: HarperCollins, 2008.

———. *The Map That Changed the World: William Smith and the Birth of Modern Geology.* New York: HarperCollins, 2001.

———. *Outposts: Journeys to the Surviving Relics of the British Empire.* New York: HarperCollins, 2003.

Yergin, Daniel. *The Quest: Energy, Security, and the Remaking of the Modern World.* New York: Penguin, 2011.

Chapter Five

Gore, Al. *Earth in the Balance: Ecology and the Human Spirit.* Boston: Houghton Mifflin, 1992.

Lean, Geoffrey, Don Hinrichsen, and Adam Markham. *Atlas of the Environment.* New York: Prentice Hall Press, 1990.

Lewis, Scott. *The Rainforest Book: How You Can Save the World's Rainforests.* Los Angeles: Living Planet Press, 1990.

Lovelock, James. *The Ages of Gaia: A Biography of Our Living Earth.* New York: Norton, 1988; New York: Bantam, 1990.

Myers, Dr. Norman. *Gaia: An Atlas of Planet Management.* New York: Anchor/Doubleday, 1984.

Revkin, Andrew. *Global Warming: Understanding the Forecast.* New York: Abbeville, 1991.

Chapter Six

Burrows, William E. *Exploring Space: Voyages in the Solar System and Beyond.* New York: Random House, 1990.

Cornell, James, *The First Stargazers: An Introduction to the Origins of Astronomy.* New York: Scribners, 1981.

Ferris, Timothy. *Coming of Age in the Milky Way.* New York: Morrow, 1988.

Hawking, Stephen. *A Brief History of Time: From the Big Bang to Black Holes.* New York: Bantam Books, 1988.

Jastrow, Robert. *Red Giants and White Dwarfs.* New York: Norton, 1990.

Trefil, James. *Meditations at Sunset: A Scientist Looks at the Sky.* New York: Scribners, 1987, 1989.

Index

About the Author

KENNETH C. DAVIS is the creator and author of the Don't Know Much About® series, and was once dubbed the "King of Knowing" by Amazon.com. He also is the author of the national bestseller *America's Hidden History* and *A Nation Rising*. Davis speaks often on radio and appears on national television and has been a guest on the *Today Show*, *CNN*, and *Fox & Friends*. He is a regular guest on NPR's syndicated *The Takeaway* and has been a commentator on *All Things Considered* and Vermont Public Radio. Davis also contributes to such national publications as the *New York Times* and *Smithsonian* and posts regularly on his website, www.dontknowmuch.com. In addition to his adult titles, he writes the Don't Know Much About® children's series published by HarperCollins. He lives in New York City with his wife.

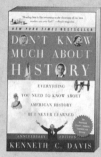
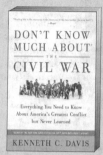

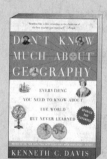
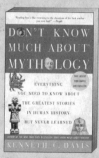

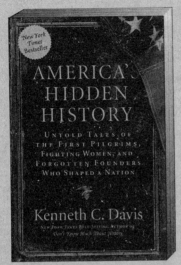
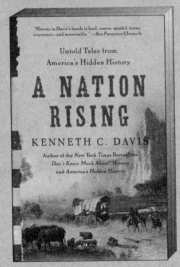